Springer
Tokyo
Berlin
Heidelberg
New York
Barcelona
Budapest
Hong Kong
London
Milan
Paris
Santa Clara
Singapore

The Origin of Competitive Strength

Fifty Years of the Auto Industry in Japan and the U.S.

AKIRA KAWAHARA

Springer

Author: Akira Kawahara
Address: 4-9-2 Mita, Minato-ku, Tokyo 108-0073, Japan
Tel.: +81-3-3451-1984 Fax: +81-3-5442-8224 e-mail: akawaha@ibm.net

ISBN-13: 978-4-431-68421-3 e-ISBN-13: 978-4-431-68419-0
DOI: 10.1007/978-4-431-68419-0
Printed on acid-free paper

SPIN:10666662

Acknowledgment

Acknowledgment

Including my years at the Automobile Section of the Japanese government's Ministry of International Trade and Industry (MITI), I have been involved in the auto industry for more than 50 years. During this period I have made many acquaintances who have directly and indirectly contributed to the development of the automobile industry, and I have learned a great deal from all of them. I have a long list of people I would like to thank, and I hope those I may have inadvertently overlooked will forgive me.

First I would like to thank the late Professor Kankuro Kaneshige (chairman of the Japan Academy of Science), who was extremely helpful to me in my student days; he was also a great source of encouragement and guidance while I was working at Toyota. I would like to mention three professors at the Production Engineering Research Institute of the University of Tokyo: the late Atsushi Watari, the late Osamu Hirao, and the late Tomoo Ishihara, who were like brothers to me. The following professors were also of great help: the late Genpachiro Konno, chief consultant to the Japan Traffic Policy Research Council; Totaro Miyashita, professor at Aoyama Gakuin University; the late Yutaka Osawa, professor emeritus at Osaka University; and Chosei Ito, professor emeritus at Sophia University.

I also express my thanks to my old friends at Tokyo Higher School, Ryoichi Nakagawa, former senior managing director of Nissan Motors; Ryoichi Kawai, honorary chairman of Komatsu Co., Ltd., who was also the director of the Machine Export Section at MITI during my tenure there; and Toyo-o Tate, former chairman of Mitsubishi Motors, who, when I asked him to intervene and get Chrysler's Lee Iacocca to stop bashing Japan, responded that Iacocca was always a gentleman when he talked to him.

I was also greatly assisted by the following gentlemen: Seigo Kakitsubo, advisor at Nitto Boseki (and director of the Automobile Section of MITI during the uproar around the so-called "people's car"); Shintaro Tabata, consultant to the Japan Science and Engineering Information Center; Genji Furusho, advisor at New Geothermal Engineering and Technology Development; and Katsuyoshi Yoshikai, director of the Council on Science, Technology and the Economy. I would also like to thank the following four individuals, who spurred me on in their research and presentation of the history of the automobile and the automobile industry: Mitsuo Nakao, former managing director of Mitsubishi Motors; Shoichi Inoue, professor at Kansai University; Yoshitake Iwasawa, former general manager of the overseas office of the Japan Automobile Manufacturers Association; and Katsunao Koiso, former general manager of the Data Office of the Japan Motor Industrial Federation Inc.

I would like to express my sincere gratitude to Toyota chairman Shoichiro Toyoda, president Hiroshi Okuda, and immediate past president Tatsuro Toyoda, for providing me with their utmost support during the writing of this text (they helped me obtain an advisory post at the Institute for International Economic Studies). I must also extend my profound thanks to members of the Institute: Bunroku Yoshino, president and director; Eiichi Kumabe, vice president; and Yoshiharu Matsumoto, standing auditor — indeed, the entire staff. Expressions of gratitude go out to Hideaki Horiuchi, chief researcher, who accompanied me on my trip to the U.S., assisted me as an interpreter, and collected many materials for me. I was also assisted by Miyako Yoda of Toyota Motor Sales, U.S.A. (TMS), in the collection and translation of reference materials in the U.S. and the compilation of collected materials by Wakako Ebata of the Institute for International Economic Studies. I also received valuable advice from Toru Sasaki, former general manager of the Business Development Division of Toyota, currently a professor at Nagoya University of Commerce and Business Administration. His advice helped me in drafting materials I used in my lectures on the auto industry at Hitotsubashi University in 1981.

In May 1995, I issued a Japanese language book, *The Origin of Competitive Strength — 50 Years of Japan & U.S. Automobile Industries* (Diamond Co., Tokyo, 1995). This book is the English version of the earlier publication. It is not, however, a direct translation. After translation, some of the original text was rewritten to be made more readily understandable for English-speaking readers. Furthermore, as two years have now passed since the publication of the Japanese version, I have, wherever possible, updated the tables and charts with newly recorded facts and figures. As the progress of the global automobile industry has been so rapid, an addendum reporting the latest progress of the industry has been included. Additionally, an author's message for the English version has been inserted in front of the main text.

Once I began making the English version I found it was quite a difficult undertaking because of the language and cultural differences. Many people volunteered to assist me, but I could not possibly have finished the work without the help of Jim Cain, Earl Quist, Jeff Smith and Martin Smith of Toyota Motor Sales, U.S.A, Inc. Along the way I encountered technical difficulties in producing the drafts, but thankfully, these were overcome through the capable assistance of Russ Bankson and Michio Inoue of TMS and my close personal friend, Nobuyoshi Nakamura.

I would like to acknowledge all those who helped me while I was at MITI and Toyota. In closing, I must thank Genichi Nakamura, professor at the West Tokyo Science University, who unceasingly encouraged me in this book, and also Tatsuo Fukuda and Eiichi Kuroki, both at Diamond Inc., who assisted me in writing and publishing the book.

I must express my hearty gratitude not only to the above-named people, but to my many friends who helped me as well. Thank you.

Akira Kawahara

Author's Message

Author's Message for English Version

Japan's auto manufacturers produced about one-half million vehicles in 1960. In the same year America's auto industry built 8 million vehicles. Who could have predicted that just 20 years later the Japanese auto industry would produce more than 11 million vehicles — 3 million more than their American counterparts manufactured in that same year — and become the world's leading auto producer for the next 10 years?

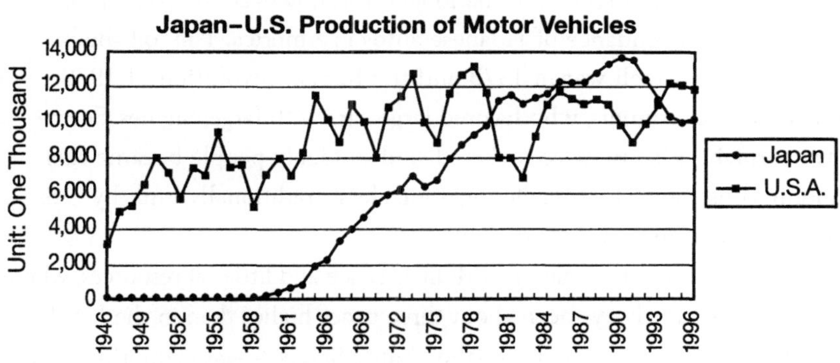

Source: Japan Automobile Manufacturing Association, American Automobile Manufacturing Association

The 1960s were pivotal years for Japanese manufacturers. They dedicated themselves as an industry to develop compact and sub-compact cars with excellent fuel economy. They succeeded in producing those vehicles at the lowest cost by adopting the "Just-In-Time" production management system which Eiji Toyoda, the honorary chairman and the late Taiichi Ohno, the executive vice-president of Toyota Motor Co. developed. At the same time, they had succeeded in developing high quality cars by introducing the statistical quality control proposed by the late W. Edwards Deming, professor of Columbia University, whose ideas have only recently been adopted by the U.S. auto industry.

In Japan not only the auto industry but many other industries followed the doctrines of Dr. Deming and from them developed the management system known as Total Quality Control (TQC). TQC enabled Japanese manufacturing industries to supply the highest-quality product at the minimum costs.

This meticulous dedication to quality was born of experience. At the end of the 1950s Japanese automakers failed to sell their cars in the U.S. market because of their poor quality and performance. By the late 1960s — after the implementation of such tools as TQC and the "Just-In-Time" production system — Japanese autos had become competitive against U.S. products and other foreign cars in quality and cost. However, the consumer perception created by the 1950s continued to adversely affect sales even after their products had been improved.

It took an unforeseen international crisis, however, to serve as the catalyst for the ascendance of Japanese autos in America. The "oil shocks" caused by Arab-Israeli war in 1973 and the Iranian revolution of 1979 surprised U.S. consumers, who favored large cars with large engines. Seemingly overnight, gas prices skyrocketed, and suddenly people became fuel economy conscious. The so-called "gas guzzlers" traditionally built by U.S. auto manufacturers lost their market.

Japan was an island nation with little space and little oil resources, where gasoline prices had always been about three times higher than in the U.S. Suddenly the world's largest market — the United States — was clamoring for the small, fuel-efficient cars which Japanese automakers manufactured in response to the demands of their domestic market. The rest was a simple matter of supply and demand. The Japanese cars flooded into the U.S. market.

Certainly no one in Japan thought their auto industry would grow so dramatically so quickly, and few in the U.S. thought their own auto manufacturers would ever be so profoundly challenged in the marketplace. To some Americans, this challenge represented much more than mere competition for car and truck buyers; it represented nothing less than a challenge to America's national interest and its leadership role in the world.

While the rapid ascendance of Japan's products threatened some in America, others sought to study how such companies as Toyota, Nissan, Honda and others had won more than 20 percent of the U.S. auto market.

Clearly many American consumers were attracted to the fuel efficiency of Japanese autos, and many stayed loyal to their Toyota or Honda because these cars proved to be reliable, high-quality vehicles. The result was a new international quality benchmark for automobiles.

Researchers from universities such as Harvard and MIT found that — contrary to the judgment of many in the U.S. — another important key to Japanese success was not low wages, but high productivity. Japan's auto manufacturers fought fiercely for market share in the highly competitive Japanese domestic market, and they sought out the new innovative products necessary to gain competitive advantage over their counterparts.

Soon top executives of the world's leading auto manufacturers visited Japan, touring Japanese plants and examining manufacturing techniques. They began to apply Japanese production methods in their manufacturing processes. Some of them — notably Toyota and General Motors — started joint production in the U.S. and automakers such as Chrysler established equity in Japanese companies.

Still, however, distrust remained. Japan continued to export vastly more vehicles to the U.S. than it imported. Even when many U.S. auto plants were obliged to close and lay off their workers due to the sudden decrease in demand for their vehicles and some of the auto manufacturers were on the verge of bankruptcy, Japan needlessly continued to supply her products as much as possible for her own sake. Japanese manufacturers who had become less competitive in their domestic market tried to find the ways and means for their survival by selling their vehicles in the U.S. market. They brought their fierce domestic competitiveness into the U.S. market. Thus the Japanese auto industry gorged on export volume, mindless of the situation of the U.S. auto industry.

Japan's government deserved its share of the blame. It was slow to remove such import barriers in the Japanese systems as tax and certification. Quite often, the Japanese government took action only when pressure was applied by the U.S. government. Only after U.S. lawmakers repeatedly advocated actions by the Japanese government to cope with the situation for almost one year, the voluntary export restriction of Japanese cars was imposed by the Japan's government in 1980.

The flames of distrust were fanned in the U.S. by some of the very persons who were benefiting from the diffusion of Japanese manufacturing techniques. Chrysler CEO Lee Iacocca — whose company imported autos and auto parts from Japan and owned a 15% stake in Mitsubishi Motors — wrote in his best-selling autobiography that Japan did not practice free trade and that it manipulated its currency to give Japanese-made products a price advantage in the international marketplace.

Iacocca's comments were perhaps even more influential because no Japanese auto executive stepped forward to rebut the Chrysler CEO's allegations. I myself told Mr. Tomio Kubo, chairman and CEO of Mitsubishi Motors, that I thought he was the best person to do so based on his company's relationship with Chrysler. Mr. Kubo, however, like other Japanese auto executives, was far too busy running his company to mount the level of effort necessary to rebut Iacocca.

Fortunately, the United States and Japan have avoided a truly damaging fight over autos. Such a fight seemed quite possible in 1995, when the U.S. threatened sanctions on Japanese luxury vehicles. The U.S. pushed Japan to agree to American-made parts purchase targets for the next several years, and Japanese government understandably balked. What was particularly problematic for the Japanese side was how inconsistent such demands were in light of the newly created World Trade Organization (WTO), which the U.S. had supported. Japan believed the threatened sanctions gave them a sound case, and they were prepared not only to take the U.S. before the WTO if sanctions were imposed but also to impose retaliatory sanctions on U.S. imports into Japan.

The two sides averted a trade war by carefully crafting an agreement giving both sides something of what each had demanded. Immediately after the conclusion of the agreement, Japanese manufacturers announced their "Global Business Plans," outlining planned purchasing activities worldwide and expansion of overseas production capacity — including the U.S.

Perhaps more importantly, however, business improved dramatically for U.S. auto manufacturers. Detroit's so-called "Big Three" — Chrysler, Ford and General Motors — were producing increasingly high-quality cars and trucks which were selling briskly.

At the same time, Japanese automakers continue to make unprecedented investments in new plants and equipment located in the U.S. All of the major Japanese auto manufacturers — Toyota, Nissan, Honda, Mitsubishi, Mazda and Isuzu — manufacture cars and/or trucks in the U.S.

Toyota, Japan's largest automaker, has made a continuing series of commitments to locate manufacturing capacity in the U.S. Soon three of four Toyotas purchased in the United States will be manufactured in North America.

Perhaps most impressively, while automakers such as Toyota were expanding manufacturing capacity in the U.S., Japanese automakers succeeded in preventing the so-called "hollowing-out" of domestic manufacturing due to production offshore by increasing domestic sales and exports to non-auto-producing countries.

Differences surely remain between Japan and the U.S. over auto trade. Most prominent is the lingering sense of American dissatisfaction over the number of dealerships selling Ford, Chrysler and GM vehicles in Japan. Understandably, the Big Three hope to have as large a network of retail outlets as possible in Japan. But contrary to some U.S. critics, the "barrier" to Big Three dealerships in Japan has nothing to do with the Japanese government nor with manufacturers, but everything to do with the specific business decisions of individual Japanese auto dealers. If Japanese dealers can profitably sell cars made by Ford, Chrysler and GM, they will. European vehicles — including those manufactured by European subsidiaries of the Big Three — have become very popular among Japanese consumers. I am confident, however, that this dispute — like those before it — will eventually be resolved amicably when the U.S. manufacturers can supply products which satisfy the Japanese customers' demands.

For the past 50 years — working for the first decade with MITI and then for Toyota — I have witnessed the growth of Japanese industry and its effect on American manufacturers and the American marketplace. Recently, I had the pleasure of reminiscing about those years with Mr. Eiji Toyoda, honorary chairman of Toyota Motor and the leader who built Toyota into the company as it is today. He said both of us were lucky enough to witness the unprecedented period in the history of the Japanese and U.S. auto industry, from the seminal ideas of General Motors' Alfred P. Sloan Jr., embodied in his book *My Years in General Motors* (Doubleday, 1963) to today's highly competitive global vehicle marketplace. It is my intention to tell the story of the Japanese and U.S. auto industries during these remarkable years and help as many people outside Japan as possible better understand how these two great institutions have competed and cooperated over this truly historic time.

June, 1997
Akira Kawahara

Contents

Contents

Preface

Preface

When the war ended on August 15, 1945, I was a naval engineering cadet at the Kure Navy Yard near Hiroshima, Japan. A week later, I was demobilized and returned to my home in Tokyo, fortunate not to find it ravaged by firebombing. At the beginning of September, a large contingent of the American occupation forces led by General Douglas MacArthur moved its base from Yokohama to Tokyo. Near my home I watched a procession of American military motor vehicles snaking along Highway 1. This truly awe-inspiring cavalcade included jeeps, two-and-a-half-ton trucks, and enormous trailers mounted with tanks and artillery. At the time, I was a 21-year-old student in the Machinery Section of Engineering at the Tokyo Imperial University. Watching that magnificent parade of military vehicles, I was more than impressed by the gap in industrial strength between Japan and the U.S. That realization led me to devote my whole life to the development of the Japanese auto industry.

I wrote a small article concerning this incident in *Nikkei Sangyo Shimbun* (one of the leading business newspapers in Japan) on May 2, 1983. The English translation of this story was carried in the July 3, 1983 edition of the *Topeka Capital-Journal* and the September 13, 1983 issue of the *Asian Wall Street Journal*. The *Topeka Capital-Journal* headline read, "MacArthur's Jeeps Were the Toyota Catalyst."

Looking for a job in 1946 was an extremely daunting task for those who had graduated amidst the confusion in the aftermath of defeat, but I was fortunate enough to join the Mitsubishi Trading Company immediately after graduation in September upon the recommendation of the Dean of Faculty, Professor Kankuro Kaneshige. However, in July 1947, Mitsubishi had to close its doors to comply with an order issued by the Allied Forces dissolving the *zaibatsu*. I decided to take this opportunity to try to fulfill my long-held dream of working in the auto industry.

An acquaintance in the Ministry of Commerce and Industry arranged an introduction to the Toyota Motor Co. by Ichibei Terasawa, director of the Machinery Division of the Nagoya Bureau of the Ministry of Commerce and Industry (later a lieutenant general in the Self-Defense Forces). Through the good graces of Mr. Terasawa, I was subsequently able to meet Shoichi Saito, a director at Toyota (later the chairman of the board) at the head office in Koromo City (currently known as Toyota City). Mr. Saito was kind enough to take me on a tour of the factories himself, but he made it clear that the company had its hands full trying to find positions for demobilized military personnel returning from the war and couldn't hire any new engineers.

Terasawa advised me that now was not the time to try to join an auto company, and if I really wanted to be involved in the industry, I should get into the Automobile Section of the Ministry of Commerce and Industry (later named the Ministry of International Trade and Industry, or MITI). As a result, a year after graduation from the university, I found myself sitting alongside new graduates taking the entrance examination for the ministry. During my entrance interview I expressed my desire to be assigned to the Automobile Section of the Machinery Division and went so far as to say I wouldn't join the ministry if I were not assigned to that section. Fortunately, I was accepted.

As soon as I resigned from Mitsubishi in September 1947, I became a technical official at MITI, receiving an appointment in the Automobile Section of the Machinery Division. During my tenure in the ministry, I turned down every regular post rotation and stayed with the Automobile Section for ten years, for career bureaucrats a record length of time in the same office. When I revealed for the first time in 1993 how I had entered the ministry to friends who had joined MITI as technical officials in 1947, they were surprised. They said they could not imagine one of their fellow officials actually wanting to work in the automotive field at that time, just after our defeat in the war.

The country was struggling through an extremely trying time in the immediate postwar period, yet we who were working in the Automobile Section of MITI were caught up in the dream of trying, in any way possible, to help establish the auto industry in Japan. Our special aim was to promote passenger car production.

After I had spent 10 years in the Automobile Section, I felt I had done all I could as a bureaucrat. In 1957 I resigned from the ministry and entered the Toyota Motor Sales Co. In learning about automobile marketing strategy at the side of the late Shotaro Kamiya, then president (known as the "God of Sales"), I was able to contribute, in my own way, to the development of Toyota. Toyota's growth has been inextricably linked to the development of the Japanese auto industry, and Japan has now become a prosperous country in which any citizen can take advantage of the convenience of the car, the symbol of 20th-century culture. I am pleased beyond measure that the dreams of my youth have been fulfilled.

In the fall of 1963, a book was published that became required reading for researchers in the auto industry as well as business managers elsewhere. It was written by Alfred P. Sloan, Jr., former chairman of General Motors, and was titled *My Years with General Motors* (Doubleday, 1963). I probably got hold of this book before anyone else in Japan. I immediately sat down to read it and soaked up every word. I had been interested in the behavior of GM, the world's number-one automaker. Indeed, my interest was so great that you could even call me a "GM watcher." I monitored GM's activities through newspapers, magazines, books, etc., not only as a task at Toyota but as a personal hobby.

This gigantic company stumbled badly in the years after Sloan left. It fell so low that at one point some analysts warned that it was in danger of falling into bankruptcy. GM has since gotten back on its feet under the leadership of Jack Smith, who took over in 1992. The lesson to be learned from GM's troubles is that the fate of even such a huge company with a formidable organization staffed by top-notch personnel depends on the quality and leadership of an extremely small number of executives at the top.

In the fall of 1993, I got together with a well-known automotive financial analyst, Maryann Keller, managing director of Furman Selz Inc., to discuss the key points underlying GM's actions. She indicated that she thought my analysis and evaluation based on the results of my years of GM-watching were on the mark. She also urged me to put my observations into book form, because she felt that a look at GM from a Japanese viewpoint would make for extremely interesting reading.

In the latter half of the 1970s, a series of research reports on the Japanese auto industry was put out in the U.S. by various researchers, including some from Harvard University. Many of these reports concluded that the Japanese auto industry had surpassed the American auto industry in terms of product quality and productivity, and further advocated learning from the Japanese. At the same time, others contended that the Japanese auto industry had pitted itself against the U.S. in an unfair manner, stealing markets, forcing some local plants to close, and throwing many people out of work. They began "Japan bashing." In many cases the resulting misunderstandings or distortions became accepted wisdom.

However, the decline of the American auto industry was brought on by the American automakers themselves. We did not use unfair means to enter the U.S, nor is it true that Japan closed its markets to foreign automobiles.

Maryann Keller authored *Collision* (Currency and Doubleday, 1993), which deals with the restructuring efforts of three automobile companies, GM, Volkswagen, and Toyota. She praises the reforms Smith, a new president, implemented at GM, and feels that the company has established itself as the most promising manufacturer for the 21st century. She rates Volkswagen second, and believes that Toyota will have trouble recovering from the double punch of a collapsed bubble economy and the strong yen.

Christopher Wood, former Tokyo bureau chief for *The Economist* and long-term resident of Japan, has written a book entitled *The End of Japan Inc.* (Simon & Schuster, 1994). While only glancing at conditions within Toyota, Wood contends that if Japanese automakers do not comprehensively rethink their hitherto successful approaches to management and product quality, they will be forced to lay off a large number of workers, thereby inviting instability within Japan.

However, there are at least three points that these authors have failed to understand. The first is the exceptional vitality and discipline that Toyota embodies. As the late Soichi Oya, a famous journalist in Japan, once noted, "Each and every worker at Toyota is prepared for battle." His observation has come to sum up Toyota's corporate culture.

The second factor is that Toyota's so-called "lean production" is a term that was not coined by Toyota itself but rather by foreign scholars and researchers. It was they who made lean production seem like a formal and fixed concept, while in reality it is extremely flexible. Thus, any evaluation of the situation based on a fixed notion of lean production is off the mark. Toyota is always changing so that it can continue to meet the challenges of a corporate environment. It adopts such measures as increasing automation, keeping down equipment and facility costs (even if labor costs a bit more), maintaining a highly efficient production line, and making equipment and facilities easier to operate by female personnel.

Third, the company possesses a product development strength that embodies the same high level of flexibility as seen in its production system. Amidst the pressures brought on by the appreciation of the yen, coupled with a worldwide drop in prices, Toyota has the ability to rapidly upgrade, develop, and produce products in line with these trends.

Foreign observers of Toyota appear not to have grasped the true nature of Toyota's corporate culture. I think that my research on GM, accompanying an introduction to the Toyota culture, will be of help in this regard. This book is a retrospective on the Japanese and U.S. auto industries during the 50 years since the end of the World War II. By comparing GM and Toyota, the biggest auto manufacturers in both countries, I have striven to show the process by which Japan's auto industry grew at an astounding rate, while during the same period its American counterpart was weakened under an oligopolistic environment and gradually lost its competitive strength.

The Japanese auto industry started from nearly nothing. The reason for its metamorphosis into a giant industry in a scant 50 years can be traced back to a point in time three weeks after the occupation of Japan, when the American occupation force allowed the production of trucks to resume, and then later in 1949, when it gave its unconditional permission to restart passenger car production. The industry was also boosted by numerous purchases of Japanese-made trucks by the U.S. military during the Korean War and the war in Vietnam and also for Japan's Self-Defense Forces. The auto manufacturers freed themselves from the quagmire of the recession that was brought on by tight monetary control measures executed in 1949 for overcoming inflation and financial debts through these vehicle procurements by the U.S. military, and were also stimulated by the booming Japanese economy (called the special procurements boom). Thus manufacturers could get the financial power needed to procure engineering expertise and set up facilities for passenger car production.

MITI's moves to protect the domestic market by restricting passenger car imports and the flow of capital were extremely important at the time. Then, after the 1960s, a parade of newcomers, ranging from makers of two-wheeled and three-wheeled motor vehicles to airplane manufacturers, entered the market, taking on the "triumvirate" of Isuzu, Nissan, and Toyota, which had been protected and nurtured by the state since before the war. A fierce battle ensued. At the time, MITI did not allow companies other than existing manufacturers to enter car production, but six out of the eight present passenger car manufacturers were postwar entrants in the market. Fierce domestic competition brought quality and productivity up to new levels and helped these companies to raise their international competitive power. It was domestic competition that, after 1980, allowed Japan to surpass the world auto giant — the United States — and become the top producer and exporter of autos in the world.

Prior to the war, the Big Three of the U.S. (GM, Ford, and Chrysler) were very powerful indeed. But around the end of the 1950s, an oligopolistic situation led by GM developed within the Big Three. This reduced competition in technology, cost, and capital investment. The Big Three stopped trying to provide their customers with better cars at lower prices and instead became exclusively concerned with turning short-term profits. It was during this period that European as well as Japanese cars began to penetrate the U.S. market. Customers' preferences shifted to small, inexpensive cars, but the Big Three were slow to pick up the shift. In the ensuing struggle to catch up, they made a series of miscalculations.

By the end of the 1970s, Japanese manufacturers had overtaken the Big Three in terms of quality and productivity. They were poised to meet the challenge presented by the oil crisis, which resulted in the sudden sales growth of fuel-frugal Japanese cars. The oil crisis served as a wake-up call for the Big Three.

In 1994 U.S. manufacturers were able to wrest back their position as top producers of autos in the world. The situation has now come full circle, with the Big Three starting to launch a full-scale assault on the Japanese market. GM's efforts have lagged behind those of Ford and Chrysler, but as of 1992 GM was on the road to recovery. Today it stands alone as the world's number-one auto manufacturer, with its nearest rival a distant second.

Around the end of the 1950s, Toyota established a foothold in the domestic small truck market, and built full-scale mass production facilities for middle-sized passenger cars. In the 1960s the company developed vehicles for the mass market, contributing to the successful motorization of Japan.

After overcoming numerous difficulties, Toyota finally succeeded in exporting its vehicles to the U.S. in the 1970s.

Toyota now holds the number-three position among the world's auto manufacturers, behind GM and Ford. It stands as a true global corporation. Toyota vehicles are produced in twenty-five major assembly plants and sold in 163 different countries or regions around the world. The source of Toyota's strength can be found in its two great leaders, Eiji Toyoda (honorary chairman of Toyota Motor Corporation) and the late Shotaro Kamiya (honorary chairman of Toyota Motor Sales Co., Ltd.). Both of them had a remarkable instinct for human potential, and encouraged many talented leaders within their companies to make their utmost efforts in the fields of engineering, production, sales, and business, while providing them with excellent leadership. Hopefully, readers of this book will recognize the extent of their achievements.

Prologue

The Japanese Automobile Industry Gets on Its Feet

MITI's Automobile Strategy in the 1950s

Until the middle of the 1930s, Japan's domestic motor vehicle industry was limited to the small-scale production of military trucks by Isuzu Motors and about 1,000 small cars and trucks named Datsun by Nissan Motors. Mid-sized and large cars and trucks, including those used by the military, were supplied by Japanese-based assembly plants run by Ford and GM.

At this time the military situation in mainland China began to heat up, and international trade relationships, especially those with the United States, started to deteriorate. The Japanese army began to stress the need to stop depending on overseas sources for automobiles and to become self-sufficient in meeting military needs. In response to this, officials of the Ministry of Commerce and Industry quickly drafted the Motor Vehicle Manufacturing Enterprise Act, which was passed through the Imperial Diet in 1936 with the backing of the military.

The Ministry of Commerce and Industry, in charge of implementing the law, gave Nissan and Toyota special assistance, supporting the setup of full-scale facilities for the mass production of trucks. (It should be noted here that in 1950 Toyota was split into two companies: Toyota Motors, which focused on production, and Toyota Motor Sales, which focused on sales. The two companies merged again in 1982, to form the present-day Toyota Motor Corporation. In this book I shall refer to the companies as Toyota, unless a distinction is required.) In the same year, Nissan purchased idle production equipment from America's Graham Paige Co. with the hope of creating a mass production system with technical assistance from the American company. Using a Chevrolet truck as a sample model, Toyota took the advanced casting technology acquired in manufacturing automated textile machines and incorporated it into the development of automobile technology. By October 1938 the company had succeeded in establishing a mass production system for trucks composed of machining, casting, pressing, and body assembling. Isuzu was already receiving army support for its production of diesel vehicles for the military, even though it was not designated under the Motor Vehicle Manufacturing Enterprise Act, unlike Nissan and Toyota.

Although Nissan and Toyota were able to create a small-scale mass production system for trucks, they needed to incorporate a large-scale American mass production system into their factories. Both companies negotiated with Ford to adopt its mass production system. Eiji Toyoda, honorary chairman of Toyota, states in his autobiography *Toyota: Fifty Years in Motion* (Kodansha International, 1987), "On December 19, 1939, an agreement was signed to establish a joint venture in Japan in which Nissan and Toyota would each invest 30% and Ford would invest 40%. This agreement was signed by Yoshisuke Ayukawa, president of Nissan, Risaburo Toyoda, president of Toyota, and Benjamin Kopp, president of Ford Japan." The plan, however, never came to be, due to surreptitious pressure from the military in the course of its preparations for the war against the United States.

Although the government was interested only in the production of trucks, both Yoshisuke Ayukawa, of Nissan, and Kiichiro Toyoda, managing director and later president of Toyota, began truck production as a starting measure. Their real target was to start the full-scale mass production of cars. Nissan moved forward at the same time to develop a midsized car to complement its ongoing production of the compact Datsun. Toyota focused on truck production while preparing for the development and production of midsized passenger cars. However, as the war intensified, these companies could no longer afford the luxury of these forays into passenger car production.

On September 25, 1945, only three weeks after the war ended and the occupation began, the Supreme Commander for the Allied Powers (SCAP) issued permission for truck production to resume. The vigor of the auto industry in Japan today can largely be traced back to the SCAP decision to allow resumption of production at such an early juncture. It made it possible for talented engineers and technicians previously involved in military production (including airplanes) to be funneled into the motor vehicle industry as soon as Japanese troops were demobilized and returned to the work force.

This permission prompted Nissan, Toyota, and Isuzu to restart their production lines, but a related memorandum also issued by the SCAP forbade the production of passenger cars. At the time, the overland transport system was completely in ruins, and the Japanese accepted the fact that the production of freight trucks had to take priority.

But the demand for passenger cars, at least in the form of taxis, expanded greatly as soon as the economy began to recover. I joined MITI in 1947. The hot topic in the Automobile Section was getting the SCAP to allow the resumption of passenger car production. I found myself traveling almost every day to see the officers in charge of automobiles in the Economic and Scientific Section at SCAP. Basically, my job was to make daily rounds and find out what kind of materials the officer-in-charge was interested in. Then I put these materials and data together and got them on his desk as quickly as possible. I soon discovered that he included this information in an auto industry report that he had to submit to his superior every week. So I started working together with a technical officer of the same section, Hikoshiro Koshida, chairman of Koshida Corporation. (We had graduated from university in the same year; he had preceded me into the ministry by a year.) We put together a weekly progress report to provide material for the internal SCAP report. I traveled with the officer-in-charge at SCAP on his inspection tours of the truck factories in a special occupation force train which had first-class sleeping cars, a dining car, and an observation car. At the time this train was the envy of the average Japanese citizen.

It was in this manner that we established close contact with members of the occupation forces. MITI as well as the automobile industry made efforts to urge SCAP to grant permission for the resumption of passenger car production. In time we began to find individuals within the occupation forces who personally understood the Japanese position, and on October 25, 1949, the long-sought permission to start passenger car production lines was issued with no conditions attached. Through this act the American army bestowed an enormous favor on the Japanese auto industry. The raw materials required in the production of autos — such as coal, electricity, iron, and steel — were in extremely short supply at the time, and a ration system was in place. There were many different reasons for restricting production of what legitimately could be called a luxury item in a vanquished Japan. We were truly grateful that, despite all the forceful arguments that could have been made against it, the Japanese auto industry was given free rein.

Receiving permission to manufacture cars did not automatically mean that a good auto could be immediately produced. However, a strong contingent made up of importers and members of the taxi industry, as well as members of economic and political circles, wanted to import foreign cars. On April 12, 1950, Hisato Ichimada, governor of the Bank of Japan, argued that the automobile manufacturing industry did not need to be fostered—a big shock to the auto industry.

The Korean War broke out on June 25 of the same year, and by the end of July the American military had placed its first orders with Japanese companies for approximately 1,300 trucks. Orders for a total of about 10,000 Japanese trucks were placed until the cessation of fighting in July 1951. We called these "Korean Special Orders." The Japanese auto industry had been going through difficulties as a result of the recession caused by the tight fiscal policies in place since 1949 and by labor union strikes. Indeed, it was standing on the edge of an abyss. This sudden demand for vehicles was manna from heaven. It allowed operations to get back on track and provided a real boost to passenger car production. Once again, the U.S. military proved instrumental in the development of the Japanese auto industry.

MITI issued its "Policy on the Handling of Foreign Passenger Car Production-Related Cooperation and Assembly Contracts" in October 1952 as one way to establish domestic passenger car production. This policy was drafted by the chief technical officer, Takaharu Mizuno of the Automobile Section. (He joined Toyota Motors in 1955, and later became chairman of Kyowa Leather Cloth Co.) Its objectives were to cope with an increase in demand for foreign auto imports while at the same time promote passenger car production. Companies able to achieve full domestic production within five years were to be designated "companies with an established capacity" and permitted to sign contracts for the assembly of foreign cars. With this designation MITI in effect limited participation in the auto industry to companies which already existed. Authorized companies were issued a monthly allotment of foreign currency equivalent to the cost of 300 completed vehicles, and if their level of domestic production increased, they were allowed to import parts commensurate with the increased production demand within their foreign currency allotment.

In December 1952, Nissan signed a technical cooperation contract for the domestic production of four-seater 1200-cc passenger cars with Austin of the U.K. Then, in February of the following year, Isuzu concluded a technical contract with the British Rootes group for the assembly and manufacture of 1100-cc Hillman Minx passenger cars, and in the same month Hino Motors signed a detailed contract with Renault, a French public corporation, covering the assembly of 750-cc passenger car 4CVs and providing for a switch to domestic production. MITI approved these technical collaborations, in accordance with its policy mentioned above. Toyota was the only company to start out on its own and to move forward without relying on foreign technology.

Thus each company strove in its own way to achieve full-scale passenger car production. Passenger car production levels stood at zero in 1950, and had reached only 800 units per month in 1953. During this period, a two-tier passenger car import system was put into effect. Beginning in July 1952, the Japanese could purchase imports through a normal route with an allocation of foreign exchange currency, or they could buy used cars from the U.S. military community. The total number of these new imports and used cars was double the level of domestic passenger car production in 1952 (Table 1).

Around the same time production of vehicles began through technical agreements with foreign companies, Toyota and Nissan began to produce vehicles domestically. They also started to improve mass production facilities. Sales of domestic models grew steadily from 20,000 units in 1955 to 32,000 units in 1956, 47,000 units in 1957, and 50,000 units in 1958. The Toyota Motomachi Plant, the first Japanese plant to concentrate exclusively on the production of passenger cars, was completed in 1959. Production that year jumped by 50% over the previous year, to 73,000 units, and full-scale passenger car mass production systems were firmly in place by 1960, when sales rose to 129,000 units.

Table 1: No. of Passenger Cars to the Domestic Market

	1951	1952	1953
Domestic cars (production)	3,611	4,837	8,789
Foreign cars Imported new cars	—	4,700	5,170
Used cars bought from Occupation Force personnel	1,650	5,476	13,467
Total supply of cars to domestic market	5,261	15,013	27,426

Source: *The History of the Japanese Automobile Industry*, Japan Automobile Manufacturing Association, 1988.

As sales of domestically produced cars climbed, it became increasingly clear that the specifications and performance of foreign cars produced domestically were not designed with Japanese conditions in mind. Most of these cars were being bought by taxi companies or corporations. The level of private ownership was extremely low. Taxis had basically disappeared from the streets during the war, and their supply was limited after the war, so foreign cars produced domestically were brought to the market en masse. These cars, however, could not withstand the brutal conditions under which they were used.

Drivers behind the wheel of what became known as kamikaze taxis would log 100,000 kilometers a year on roads that were in horrendous condition (only 8% of the roads were paved at this time). The prevailing attitude among taxi drivers was that it took them two years to recoup the purchase price of their cab, and the next 100,000 kilometers logged during the third year was profit. Since these drivers were purchasing new vehicles every three years, they were extremely important customers for the automakers. Unfortunately, it became clear that the Austin, Hillman Minx, and Renault 4CV did not have the required durability. These autos satisfied the goal of complete domestic production during the target period, but were gradually withdrawn. It is thus difficult to characterize the technical cooperation with foreign manufacturers as real success, but Nissan was said to attain valuable access to car body manufacturing and painting technology.

The Toyota passenger car at the time was a scaled-back version of a small truck chassis over which a cab was mounted. It was tough and able to withstand the rigors of the roads, but a smooth ride was not one of its features. In January 1955, the Toyopet Crown was put on the market. This car was equipped with front-wheel coil spring independent suspension and a revolutionary three-layered leaf-spring-type suspension system for the rear wheels, providing a greatly improved ride. Although this vehicle fell short in running performance as compared with foreign cars, it was far superior to domestically produced foreign cars when it came to the toughness required to meet Japanese road conditions.

MITI's auto industry policies in the 1950s can be summed up as follows: The "Big Five" were given preference in the promotion of postwar passenger car production. They were made up of the "triumvirate" (three companies designated under the prewar Motor Vehicle Manufacturing Enterprise Act, i.e., Nissan, Toyota, and Isuzu, which supplied the military), Hino Motors (an offspring of Isuzu), and Kawasaki Plant of Mitsubishi Heavy Industries. Hino and Kawasaki both manufactured military vehicles during the war. Manufacturers of airplanes, construction machinery, machine tools, or textile looms were shut out of the car manufacturing sector. Furthermore, imports of foreign vehicles and the inflow of foreign capital were kept to a minimum until the production of passenger cars was firmly established.

The preferential treatment given to the original triumvirate, later expanded to the Big Five, was at the core of MITI's postwar auto industry policy. Given the need to establish a source for midsized trucks, which were critical to the recovery of overland transport, it is easy to understand the wisdom of concentrating government assistance measures on the triumvirate, which had been involved in production during the war, and on the two companies which had diesel engine production facilities for the manufacture of large trucks. It is much more difficult, however, to argue the merits of a policy that gave preferential treatment to the Big Five and excluded other manufacturers from entering passenger car production. This is particularly true in light of the fact that none of these companies had experience in full-scale production of passenger cars. Today, a total of eight companies manufacture passenger cars in Japan. Six of these companies were initially kept out by the MITI policy. I believe that if a policy based on free competition had been introduced, Japanese passenger cars would have had the strength to take on the European and American markets five years earlier, or by the end of the 1960s.

One reason auto policy moved in this direction was that both Shinsuke Kishi, director of the Bureau of Engineering in the prewar Ministry of Commerce and Industry, and Yoshiteru Kogane, director of the Industry Policy Section in the same ministry, who had drafted and promoted the Motor Vehicle Manufacturing Enterprise Act, entered politics after the war. As important politicians they exerted considerable influence on postwar MITI policy. Kishi served as prime minister from February 1957 to July 1960, while Kogane formed the Automobile Diet Member Association and acted as its chairman, controlling MITI's automobile policy. Many MITI bureaucrats already had experience controlling the economy during the war under these two men. They thus implemented this new policy without questioning its inherent inefficiency or the preferential treatment afforded the Big Five.

Rise of Independent Two-Wheeled and Three-Wheeled Vehicle Manufacturers

Plants manufacturing uniquely Japanese three-wheeled trucks, independent of the Big Five, began popping up in 1946. These included Matsuda, Daihatsu, and Nihon Internal Combustion, which had been active in the field even before the war, as well as plants involved in airplane and machine tool manufacturing during the war, which were subsequently converted to civilian production. Prewar motorcycle production was limited to three companies, Rikuo, Meguro, and Miyata, but new participants (especially airplane manufacturers) emerged in the field after the war. While these companies did not receive any special support from MITI, the scope of their activities was truly startling. Three-wheeled truck production consistently outstripped the production of four-wheeled trucks in the postwar period up to 1957. In one year during this period production was more than 100,000 units. It is even reasonable to say that these manufacturers took the lead over the Big Five and built the foundation for the postwar Japanese auto industry.

When I entered the ministry, Hikoshiro Koshida, who had preceded me by one year, was placed in charge of the Big Five, and I found myself in charge of the other manufacturers. In October 1947 I was dispatched to observe running tests of three-wheeled vehicles (these tests had been entrusted to the Society of Automobile Engineers of Japan). I ended up riding in the back of a three-wheeled truck for three days and two nights, bouncing over the war-ravaged, pockmarked National Highway 1, which runs 500 kilometers from Osaka to Tokyo.

As I remember, the companies participating in this test were Daihatsu, Matsuda, Nihon Internal Combustion, Mitsui Seiki, Aichi Kigyo (a manufacturer of airplanes during the war, currently known as Aichi Machine Industry Co.), Meiwa Kogyo (a manufacturer of large hydroplanes during the war, currently known as Shin-Meiwa Industry Co., Ltd.), and Mizushima Plant of Mitsubishi Heavy Industries. The three-wheeled trucks had 500-cc single-cylinder air-cooled engines (with the exception of one that was water-cooled) and could carry a 500-kg load. We drove practically the entire route in the rain. Making it through Suzuka Pass, Sayono Nakayama, and Hakone Pass was not easy. Though the run was plagued by trouble, all vehicles completed the journey, largely because of the enthusiasm and earnestness of the engineers and mechanics involved.

Sales of three-wheeled trucks subsequently boomed. These vehicles played a major role in the transport of small loads, but the maximum load limit of 500 kg proved to be too much of a restriction. This problem spurred the Ministry of Transport, the National Police Agency, and the Ministry of Home Affairs to work with business on the revision of related laws and regulations to make it possible to manufacture and market stronger and larger three-wheeled trucks. By the latter half of the 1950s, 1200-cc long-deck two-ton trucks had appeared on the market. Before the Big Five got serious about producing small four-wheeled trucks, the three-wheeled trucks contributed dramatically to ensuring that Japan could transport small loads in the post-war period.

I was also placed in charge of the two-wheeled-vehicle industry. This industry got its start by putting an engine on a bicycle. I visited Soichiro Honda of Honda Motors right after the company's start-up of full-scale motorcycle production. There he was in the plant, dressed in a production line worker's uniform, tinkering with a fancy fiberglass model of a scooter. This scooter would later be put on the market under the name Juno. He spoke with me passionately about his aspirations for the future, remarking, "Right now I'm driving a Jaguar and until recently I was driving a Porsche. I want to build a car that is as good as those cars." I was struck by his drive and passion, and I thought that this was a man going places.

My major task working with two-wheeled vehicles was setting up size classifications for a range of laws and regulations in order to ensure conformity with standards for two-wheeled vehicles produced in developed countries like Germany, Great Britain, France, and Italy. I visited Masaharu Gotoda, director of the Accounting Section of the National Police Agency (currently a member of the House of the Representatives and a former deputy prime minister), with Hideichi Takamine, the director of the Automobile Section, who had attended Himeji High School with Gotoda. With his input we resolved what was thought to be the most difficult area, revision of the drivers' license system. These revisions made it possible for Japanese manufacturers to produce engines corresponding to the European classifications of 50, 125, 250, and 500 cc. I suppose I was influential in the subsequent participation of Japanese motorcycles in international races and in boosting sales to the European export market. Annual production of two-wheeled vehicles exceeded 1 million units in 1960 and 2 million units in 1964. Actually, by the beginning of the 1960s Japan had achieved worldwide domination of this market.

A total of 11 automobile manufacturers are active in Japan today, and with the exception of the Big Five, all grew from the companies producing two-wheeled and three-wheeled vehicles. Only two of the eight passenger car manufacturers in the market today are former members of the triumvirate or Big Five. The Kawasaki Plant of Mitsubishi Heavy Industries was a member of the Big Five, but the passenger cars put out by Mitsubishi Motors are produced by the Mizushima Plant and Nagoya Plant of Mitsubishi Heavy Industries. These two plants can be traced back to a company which was originally involved in making two-wheeled and three-wheeled vehicles, long before the mass production of four-wheeled vehicles started. The dramatic growth of two-wheeled and three-wheeled vehicle manufacturers contributed greatly to the establishment of the automobile parts and components industry in Japan.

The Shelving of the "People's Car" Plan

American soldiers based in Japan around the end of the 1940s bought Fords, Chevrolets, and other American cars at the PX. They would polish them up, find a pretty girl, and go for a drive. Watching these soldiers made me wonder why anyone born American could own a car, the symbol of 20th-century civilization, but if you happened to be born Japanese, a car was beyond your reach. We needed to do something to rectify the situation. The price of cars produced as a result of MITI's policy of nurturing passenger car businesses was more than twice the yearly income of the average citizen. As I mentioned earlier, most autos were used as taxis or by companies; they were far beyond the reach of the individual consumer.

In the 1950s, a foreign currency framework was set up for the import of motor vehicles, and some foreign cars gradually began entering the market. MITI allocated foreign currency to companies evenly, regardless of manufacturing country or type of car, as a way of preventing a particular auto or auto company from establishing too strong a market base. As a result, almost all car models manufactured in the world at the time were imported into Japan. As a bureaucrat I availed myself of the opportunity to test-drive almost all the cars that were shipped in. In fact, one other official in the Ministry of Transport and I were probably able to test-drive more different types of cars than anyone else in Japan, if not the world. The vehicles that attracted my attention the most were the Citroen 2CV and the Fiat 500.

In 1950, Ryuji Tomiya of Suminoe Works Co. produced a prototype of a two-seater 360-cc passenger car, the Flying Feather. I hardly remember my first meeting with Tomiya, but according to him, he submitted a report to MITI on the completed prototype but couldn't get us to give him the time of day. He then turned to Tadamasa Yoshiki, director of the Society of Automobile Engineers of Japan, an acquaintance since his Nissan days, and asked him to mediate. I was subsequently sent by the ministry for an inspection. The car in question was a stylish, compact version of the MG and had the personal backing of University of Tokyo professors Atsushi Watari and Osamu Hirao, Professor Emeritus Kiyoshi Tomizuka, and section manager Yutaka Katayama of Nissan (who later became president of Nissan U.S.A.).

The Flying Feather went through several prototypes and was put on the market in 1954. It weighed 400 kg and had an air-cooled four-cycle 360-cc engine and a four-wheel independent suspension (this last feature was its main attraction). It received a warm reception at the first Tokyo Motor Show, held in the fall of that year. However, right about the time that Suminoe Works embarked on production of the Flying Feather, it lost its subcontracting car body work from its parent company, Nissan, and this, together with the burden of capital investment for its fledgling vehicle production division, drove it into bankruptcy after only about 200 units had been manufactured. One of these cars is currently on display in the Toyota Automobile Museum on the outskirts of Nagoya.

With the Flying Feather, Citroen 2CV, and Fiat 500 in mind, I wrote an article entitled "A Concept for a New People's Car" for the February 1953 issue of the magazine *Ryusenkei*. The thrust of my article was that Europe's approach to the car and its production trends had changed, even though Europe still lagged behind the United States in terms of the spread of the automobile among the general populace. The new approach vilified the notion that the auto is the symbol of the egotism of the wealthy, emphasizing instead that cars were so convenient that they should be available to all in the twentieth century. Among the cars which embodied these trends were the Volkswagen, Renault 4CV, and Citroen 2CV. The Volkswagen (which literally means, of course, "people's car") was designed to bring the car to the masses. It thus seemed that we in Japan should also be working to bring the benefits of the 20th century to the common folk, getting them behind the wheel.

At that time there were approximately 200,000 people making more than 600,000 yen a year. Using this income level as a starting point for promoting the spread of the auto, I believed that these people should be able to buy a car for about 200,000 yen, or one-third of their annual income, based on the index of car-purchasing capability in the U.S. I reasoned that if a car weighed 400 kg or less and had a 500-cc air-cooled two-cylinder engine, it could be manufactured and sold at that price.

I tried out variations of this argument in different venues and was able to persuade a considerable number of people of its soundness. One morning in March 1955, approximately two years after I had first presented my paper, I was called in by Seigo Kakitsubo, director of the Automobile Section. He told me to put together a "General Plan for Promoting a Car for the People" based on the People's Car Plan that I had been championing. In later years, Kakitsubo stated that the People's Car Plan helped give shape to the passenger car industry.

MITI announced the plan in May 1955. The objective was to provide government support for the development of a car that could be mass-produced at a price within the reach of the general populace. The proposal can be summarized as follows:

- The car should be able to reach a maximum speed of at least 100 km/hr and travel at least 30 kilometers on a liter of gas at 60 km/hr.
- It should seat two to four passengers and handle a load of at least 100 kg.
- The engine should have a displacement of 350 cc to 500 cc.
- Monthly production should be targeted at 2,000 units at a manufacturing base price of 150,000 yen or less. (This figure was later revised to a sales price of 250,000 yen.)
- Each company interested in producing the vehicle was to manufacture a prototype by a set date, ensuring it met these specifications.
- MITI would test these prototypes and select one suitable for mass production.
- MITI would invest capital for production of the selected car and support its manufacture.

This proposal sparked a huge debate not only within the auto industry but also among academic circles and opinion leaders. The auto industry, which was composed of the Big Five, opposed any plan which gave companies other than the Big Five a chance to participate in passenger car production. Genshichi Asahara, chairman of the Automobile Manufacturers Association and president of Nissan Motors, was vehemently opposed, saying, "Not only is it impossible to build a car at such a low price, but it isn't even necessary. If someone wants a low-priced car, they can buy a used one." He did not accept the principle behind the proposal of offering superior automobiles at a low price so that more Japanese could experience the convenience of owning a car. The Automobile Manufacturers Association filed a formal objection to the plan, and MITI decided that its implementation was premature.

Chairman Asahara later let it be known that while he could not, as chairman of the Automobile Manufacturers Association, agree with the People's Car Plan, he personally thought it was right on the mark (see *Chronicles of the Japanese Automobile Industry History*, edited and published by the Japan Motor Industrial Federation).

Recognition of the Need to Develop a Car for the Masses

I had anticipated a negative reaction to the People's Car Plan, and as its author I had a second aim. The law gave mini-vehicles special consideration in areas related to drivers' licenses, commodity tax, possession tax, and registration. However, three-wheeled and four-wheeled vehicles had to have an engine of 360 cc or less to be classified as mini-vehicles, and there were also restrictions on total length and width. I felt that an engine of 360 cc or less would not meet the challenge of the expressways of the future and that the vertical and horizontal dimensions were too confining to seat three to four passengers. Once the specifications for the car of the people had been set, my plan was to call for a change in related laws and regulations. Even if the production plans were not approved, my aim was to at least have the limitations on external automobile dimensions and engine displacement revised. However, objections were raised from within the auto industry, and for some time no revisions were made.

Up to that time, no manufacturer had embarked on production taking advantage of the legal incentives for making four-wheeled mini-vehicles. However, two-wheeled and three-wheeled vehicle manufacturers other than the Big Five decided to take advantage of the debate and embarked on the production and sales of four-wheeled mini-vehicles. These companies included Matsuda, Daihatsu, Mitsubishi (then known as Naka Nihon Heavy Industries), Fuji Heavy Industries, Honda, and Suzuki. The production of four-wheeled mini-vehicle passenger cars stood at zero in 1955 but climbed to 36,000 units in 1960, 78,000 in 1963, and 120,000 in 1966. The production of four-wheeled mini-trucks exceeded that of passenger cars and reached 472,000 units in 1966. Based on the success of four-wheeled mini-truck production, these companies launched into the production of compact passenger cars, a market that MITI had hitherto worked to keep them out of.

Toyota was the only one of the Big Five to aggressively move into the production and development of a car for the general public. One day, at the beginning of 1956, while the People's Car Plan debate was raging, Hanji Umehara, director and general manager of the Engineering Department (later managing director), and section manager Tozo Yabuta (later managing director and president of Aichi Steel Works) of Toyota came to my office to report on their development of a prototype of a car for the general public. I was truly happy to discover that despite the fact that it was clear that the People's Car Plan had failed, a manufacturer was trying to develop a car for the masses. In the fall of 1956, I was invited to a Toyota plant and asked to test-drive their first prototype. It had front-wheel drive and an air-cooled horizontally opposed two-cylinder 700-cc engine. A compact, it had a length of 3.65 meters, a width of 1.42 meters, and a height of 1.4 meters, yet the front seat was roomy. Perhaps because it was their first prototype, the steering wheel was a little heavy. (I have recently discovered that research and manufacture of this prototype started around 1954 under the direction of Eiji Toyoda, then senior managing director.)

Mitsubishi also seized the opportunity presented by the People's Car Plan by developing a car which subsequently sold as the Mitsubishi 500. Mitsubishi's top management, however, did not show the same interest in the project as top management did at Toyota. It was reported that Yoshinari Kawai, president of Komatsu Co. (a manufacturer of construction equipment), looked into a plan to develop this type of car with the help of Porsche. Nissan Motors did not readily understand the need for a car for the general public, reasoning that since they already had the Datsun passenger car, why make an even smaller one? While the Datsun was clearly aimed at the general public in the prewar period, after the war it was bought mainly by taxi companies and corporations, and its specifications and structure no longer fit the needs of the individual. Nissan didn't seem to mind that the Datsun was priced out of the range of the average consumer. It was not until after Toyota introduced the Publica that Nissan finally got started on a prototype for a similar vehicle, but Toyota is still ahead of Nissan to this day.

The First Definition of Productivity, and the Difference Between Japan and the U.S.

The Japan Productivity Headquarters was founded by the Japanese financial world, with capital assistance from the U.S. from 1955 to 1962, to promote productivity improvement in Japan. Establishment of this body was prompted by the U.S. government in order to provide opportunities for on-site study of the superior productivity of American industry. Before this, the U.S. had helped reconstruct England, France, and West Germany via the Economic Cooperation Act (1948), on which the Marshall Plan was based. The program in Japan was similar.

Study groups from various industrial fields were dispatched to the United States to get a firsthand look at American industry's superior productivity and its underlying management techniques. The Japanese auto industry sent executives from the Big Five as well as representatives from the automotive parts industry and labor unions. These groups visited automakers such as GM, Ford, Chrysler, Studebaker, and Kaiser-Frazer and equipment manufacturers like Cincinnati Milling Machine and Gleason. The Japanese teams were taken under the wing of all of these companies, given extensive tours of the factories, and tutored on all aspects of American management techniques. The information that these travelers took home contributed greatly to the later development of the Japanese auto industry.

Around 1954, MITI received information that the productivity improvement movement had produced good results in Europe as well. The first thing MITI bureaucrats looked at was the true meaning of productivity. To produce something, you need people and equipment—or, to say it another way, labor and capital. Productivity is measured by the efficiency with which one invests these two things to increase the value of what is produced. Atsuo Nakajima (at that time working in the Casting Product Section in the Bureau of Heavy Industry at MITI and later general manager at Nippon Steel Corporation, then managing director at the Space Environment Utilization Center) and I defined productivity using the following two equations:

Labor productivity = added value/number of workers

Capital productivity = added value/tangible fixed assets

If productivity is so defined, then the following equations can be arrived at:

Labor productivity = (added value/sales volume) x (sales volume/number of workers) = added value ratio x sales volume per worker

Capital productivity = (added value/sales volume) x (sales volume/fixed tangible assets) = added value ratio x tangible fixed asset turnover ratio

If tangible fixed assets per worker are defined as the labor-equipment ratio, then the following relationships can be obtained:

Capital productivity = labor productivity/labor equipment ratio

Labor productivity = capital productivity x labor equipment ratio = added value ratio x tangible fixed asset turnover ratio x labor equipment ratio

Today these definitions and related formulas are in widespread use not only in Japan but also in the United States.

With the backing of Ryuzo Sato, an officer in the Automobile Section (and currently chairman of Nippatsu Seimitsu Co.), I worked with Nakajima using these formulas to compare the productivity of the Japanese auto industry with that of the Americans and Europeans. We tried to predict what level of development the Japanese auto industry could attain over the approximately eight years up to 1963. We also studied how to rationalize related industries, and what measures were essential to the advanced development of the Japanese auto industry. The results of this work were summarized in our book *The Future of the Japanese Auto Industry: Conditions for Advanced Development*, which was ready in the summer of 1956. This book set forth recommendations for the directions in which the Japanese auto industry should move. However, it took a year from the time the manuscript was submitted to the publisher before we could get it published, due to financial problems within the company. It finally came out in June 1957. At the time there were few readers interested in the auto industry, and the publishing company ended up with a substantial number of unsold copies. Then, using the name of the Bureau of

Heavy Industry of MITI as the editor of the second edition, the publisher sold directly to relevant companies. Today this second edition can be found in various places, and the Heavy Industry Division is listed as the author in auto-related reference lists. In reality, of course, it was written by the three of us on our own time, after office hours. Our definitions of productivity were used in an economic white paper written by the Economic Planning Agency, and they have come into common use in economic circles.

The labor productivity of Japanese auto companies was one-fifth that of GM, while capital productivity was about one-tenth. These figures illustrate the quantitative differences which existed between Japanese and American productivity. Labor productivity was one-half and capital productivity was about three-fourths that of Renault in Europe. We emphasized the substantial room for improvement through such measures as the two-shift working system employed in leading auto companies and the introduction of automation equipment like transfer machines. Furthermore, we said, Japan needed upgraded technology for its auto parts manufacturing, press processing, machine tools manufacturing, and the processing of raw materials. We concluded that the Japanese auto industry could be on a par with the European industry in about eight years.

In 1955, MITI commissioned the Japan Auto Parts Industry Association (JAPIA) to conduct a survey of 338 parts manufacturers. I directly oversaw the project, which involved collecting information, analyzing productivity with IBM computers, and editing the report. The computer analysis was an enormous undertaking, and we used 140,000 punch cards. From the JAPIA I had the valuable services of Yasusada Nobumoto, managing director of Akebono Brakes Co. (currently chairman). He helped tremendously in writing a report, which was finished in March 1956. The main text was 220 pages, with accompanying reference materials running to 1,530 pages. Printing and distribution were not completed until June 1957.

Once the Machinery Industry Promotion Act of 1956 was promulgated and enacted in June 1956, automobile parts were classified as part of the special machinery industry. This designation meant that a company with a suitable productivity improvement plan could qualify for financial assistance. Our survey served as a basis for these improvement plans.

I submitted my resignation to MITI at the end of July 1957 and joined Toyota Motor Sales. The decade ending in 1957 was an important period during which the direction of the postwar Japanese automobile industry had been decided. I am satisfied today that I was able to contribute to the establishment of the car industry in my role as an official heavily involved with formulating the government's automobile industry policy. During my stay at MITI I was able to engage in discussion with some of Japan's top economists at the Economy Research Institute at MITI, including Tadao Uchida, Kiyoshi Kojima, Kimio Miyagawa, and Miyohei Shinohara. Not only was I the recipient of much kindness at the ministry, but I also gained some knowledge of macroeconomics that was of considerable use to me in my later activities.

Chapter I

Toyota Moves into the Fast Lane

Toyota Moves into the Fast Lane

I-1. The Big Time — Launching Toyota's American Operations

I entered Toyota Motor Sales Company (hereafter referred to as Toyota Motor Sales) in July 1957 as section manager of the newly established market research section in the Export Department. Prior to my joining the company, the president, Shotaro Kamiya, had gone on a market research tour in North, Central, and South America from the end of March through the middle of May 1957. In the course of his travels, he decided to move immediately into the U.S. market. My job was to make the necessary preparations for this new project.

In August 1957, we shipped two Toyopet Crowns to Los Angeles as sample cars. The late Seishi Kato, senior managing director (who became president and chairman), conducted on-site market surveys using these cars. He reported back to us that they were well received. He estimated that sales of at least 2,000 units a year were possible. President Kamiya then drew up a plan for moving into the U.S. market, based on his own on-site experiences and Kato's report.

Kamiya felt the performance and quality of the Toyopet Crown still needed improvement, but if we satisfied our customers through exemplary service, we should make steady inroads into the American market. His plan was to have Toyota Motor Co. (hereafter referred to as Toyota Motor) develop a full-scale export model to be released in the future. To get these cars to the market, he said we should not rely on independent importers but rather establish our own import sales company in Los Angeles, owned 100% by Toyota (50% each by Toyota Motor and Toyota Motor Sales). This company would handle wholesale and retail sales, monitor the American auto market, and establish a foothold for future advances.

Toyota applied to the Japanese government for permission to make an overseas investment of $1 million. Approval was granted in the fall of 1957. $750,000 was invested in cash, and the remaining $250,000 was in commodity, composed of spare parts. Toyota Motor Sales, U.S.A., Inc. (hereafter referred to as TMS) was founded in October 1957, with president Kamiya also serving as its president.

When capital arrangements were more or less finished in Japan, I was dispatched to Los Angeles at the end of 1957 to help with the preparation for sales. It was my first trip abroad.

Shoji Hattori (a Japanese-American graduate of the University of California at Los Angeles [UCLA], deputy general manager of the Export Department of Toyota Motor Sales), and Hiroshi Kamiya (eldest son of president Kamiya, currently chairman of Nanboku Co., Ltd.) had preceded me there. Two weeks after my arrival, Junkichi Goto was appointed general manager of general affairs, accounting, and finances (he later became senior managing director at Indonesia Oil Co.), while Toshiro Tajima was appointed service general manager. Approximately one month later, Tokutaro Kobayashi, former Oil Department general manager, was appointed vice president.

It took about one month to study local procedures, from importing autos to retail sales. I couldn't find anyone well versed in all aspects of these procedures, so I visited banks, importers, auto sales outlets, and government agencies to learn as much as I could about their area of expertise. I then took this information and charted the flow of goods and funds in detail, put together the necessary documentation, and took the survey results to the general manager of the Automotive Department at the Bank of America for his comments. He praised the thoroughness of the report and was surprised by the quality of our survey.

One day, looking through a telephone directory for listings of auto importers such as Volkswagen and Volvo, I noticed that in every case import, wholesale, and retail divisions were independent companies. I immediately consulted a lawyer. He said that companies were required to conduct "arm's-length transactions," even with their own subsidiaries. The concept behind arm's-length transactions is based on equal conditions for every company—subsidiaries are not to receive special treatment. Furthermore, compliance with this rule should be verifiable throughout the organization. An excise tax is assessed on the sales price from the importer to the wholesaler, and this must be verified to be a fair market price. Similarly, it must also be possible to ascertain that the sales price paid by the subsidiary's retailer to the wholesaler is a fair price. Companies then divide themselves into three, to show that they sell their goods at the same price and under the same conditions to outside capitalized companies as they do to their own related companies.

We realized we would have to adopt the same approach. TMS financed the establishment of two separate companies: Toyota Motor Distributor and Hollywood Toyota. In Japan the import customs tax is calculated by multiplying the customs tax rate by the CIF price. In the United States, if it is determined that the Japanese factory shipping price is a fair price, as determined by the absence of a difference in price between Japanese domestic models and export models (or a difference in price attributable to such factors as fittings, specifications, or incentives), then the customs duty is calculated by multiplying the price of the vehicle by the customs rate. A manufacturing cost calculation report must be submitted to verify that a product is not being dumped on the market. We were taken aback when we discovered that materials which are normally considered to be confidential and not submitted to any agency in Japan were demanded by the American authorities.

At the time, almost all our communications with our home office were by mail because it was extremely difficult to communicate by telephone for a variety of reasons, such as poor audio quality (those were the days of shortwave transmission), the time between placing a call and connecting with the opposite party, which could take hours, and the high cost. What today is a simple matter was then essentially an exercise in frustration. Telegrams had to be written in Roman letters, and every time we wanted to send one we had to go to the telegraph office. Moreover, reading an entire Japanese document in Roman letters was a real challenge. It took quite a bit of explanation to get the head office to understand our situation and get their cooperation. Airmail took six days; even if the Japanese side responded immediately, it took about two weeks to get the response of the head office. I would spend most of the work week trying to get everything ready for sales and many of my Saturdays and Sundays writing letters to the head office detailing what was going on.

The next step was to get contracts signed between importers and wholesalers and between wholesalers and retailers. We had to satisfy the conditions for fair trade and ensure the profit margins of manufacturers and dealers. Fortunately, a lawyer who worked for Volkswagen was willing to take us on as clients, and we came up with a contract that fitted the needs of the time perfectly. It was through these activities that we came to understand that all business in the United States had to be based on the notion of fair trade.

The next item on our agenda was to obtain certification from the California Highway Patrol. (The American TV drama *Highway Patrol* was popular in Japan.) The patrol, an agency of the California State Department of Motor Vehicles, oversees the state vehicle code, which regulates both vehicle transport and traffic laws.

The turn signal on the Crown was a cross arm extending from the center pillar of the car body to the outside, as was common in British vehicles. Under the California vehicle code, the signal had to be equipped with a flashing light. The device had to be taken to the Highway Patrol and be tested for durability. At that time in California, drivers signaled turns by sticking an arm out the window, so I argued that a cross arm indicator should be acceptable. The official rejected this, saying that a design could never have "too much safety."

Standards governing headlight brightness were based on the sealed-beam lamp, which was widespread in the United States then. Since we were still at the prototype stage in Japan, we imported our cars without headlamps and purchased GE headlamps stateside. The quality of Japanese-made windshield wipers was also not good enough and their patents were not clear, so we purchased Trico-made wipers in the U.S. Regulations called for reflectors on the rear of the car with standard degrees of reflection. Since it was impossible to procure Japanese products that would conform to these standards, we bought American reflectors. The tires were made by Dunlop Japan because Dunlop U.S.A. had a nationwide repair network and could guarantee quality.

Of course we also changed the design to left-hand steering, though it was not mandatory under the law. The issue of right-hand steering has recently become a source of controversy in the export of American cars into Japan. It should be noted that, like American law, Japanese law does not require that steering conform to the side of the road the nation drives on.

Note: The Toyopet Crown was a midsized auto manufactured entirely in Japan with a chassis designed specially for a passenger car. The car was introduced into the market in January 1955 and was the first passenger car manufactured in Japan using mass production facilities. The Crown immediately proved popular for corporations and taxi companies. Production costs dropped as output and sales rose, and these savings were passed on to the customer through a series of substantial price cuts from December 1955 through August 1956. The vehicle even penetrated the small-business market. Yutaka Tsuji, a reporter with the newspaper *Asahi Shimbun*, traveled 50,000 kilometers in a Crown from London to Tokyo. It took him eight months, from April to December 1956. The Crown also completed a road rally along the perimeter of Australia in the summer of 1957. One of the drivers was Yukiyasu Togo, who later became president of TMS. These results proved that the car could, to a certain degree, show durability in difficult road conditions. However, there were no freeways in Japan at the time where one could cruise at 100 km, and it turned out that the car performed poorly at high speeds. It weighed 1.2 tons, so it was too heavy for its 1500-cc engine. We came up with this figure by comparing the Crown with the 930-kg Opel Record put out in West Germany. President Kamiya asked Toyota Motor to develop a serious export model, and the Crown was put on the export market first to establish a foothold in the United States, particularly California, which would serve as a base for the development of a car especially designed for export. Those two Crowns represented Toyota's first export model in the United States.

I-2. Fifteen Long Years Establishing an American Base

President Kamiya wanted to entrust the management of TMS to a top-notch local manager. This matter was handled by the late On Nakae (in charge of exports, later appointed executive vice president, he came over about once a month from Japan) and deputy general manager Hattori. They selected Eric Hansen of Volvo Import Co. and gave him total authority over marketing management of the company. Hansen, an MBA from the University of Michigan, was young and bursting with ambition. After a period of time, Harold Johnson, who had served as service manager of the Edsel Division of Ford, was appointed service manager.

The American staff was gradually assembled under Hansen. Most of the staff had previously worked for European import companies. They jumped right into establishing dealerships, but few car dealers showed any interest in a completely unknown, nondescript Japanese car. GM and Ford hardly gave us a passing glance. As I remember, by the time we opened for business we had established relationships with only six or seven outlets. Many of these were used car dealers.

We finally opened for business on July 10, 1958. Our first customer at Hollywood Toyota was a refined, elderly lady. I turned over the keys to the car (with a few commemorative items) at a designated rendezvous point. I remember all that I could do was pray the Crown wouldn't break down. Since we never received any complaints from her, I can only assume she was a careful driver who did not speed along the freeway. At the time, many women and senior citizens avoided expressways.

In truth, we were under a great deal of pressure. On the freeway, at speeds of around 100 km, the Crown's weak point became painfully clear— the car body and chassis would begin to vibrate violently. If the car was driven at high speeds for more than several thousand kilometers, the bracket supporting the engine generator broke due to metal fatigue. Executive vice president Kobayashi and general service manager Tajima of TMS were the victims of just such a breakdown traveling through the Arizona desert. They had to spend a cold night on the side of the road until they were picked up by the highway patrol the next morning. It was a serious problem. We would run out of replacement parts and couldn't get parts air-freighted on time. We had a thicker and sturdier bracket made in Los Angeles, but it still wasn't good enough. When the cars were driven at high speeds, oil would leak from the engine and the differential gears, creating many disgruntled customers.

Unfortunately, as we continued to find more defects and problems, the Toyopet got a reputation as a lemon, and sales languished. TMS soon found itself locked in a life-or-death struggle.

In the spring of 1959, we assembled a number of major cars from different countries and conducted test drives through the Los Angeles suburbs. We quickly found that the best-selling import, Volkswagen, and the popular Renault Dauphine did not stand up well to crosswinds. They also lacked power and had poor braking. The only car that could be given good rating was the Peugeot 403. We had the impression that in the near future we would be able to overcome the technical gap between these autos and our own as far as performance and quality were concerned.

Finally, at the end of 1959, the prototype of our long-awaited export model arrived. The car incorporated a revolutionary concept: The front-wheel suspension unit had a torsion bar, to make the car lighter, while the rear unit had an exclusive four-link cantilever device which combined a leaf spring with a coil spring. In contrast to the Crown, it could be driven at high speeds on the open road. However, we were unsure of the strength of the chassis and the car's maneuverability, so we put it through extremely rigorous driving tests to discover any weak points. We drove the cars so hard that Kenya Nakamura, chief engineer in charge of development of the Crown, came to the conclusion that everyone who worked at TMS was a maniac behind the wheel.

This car, dubbed the Tiara, represented a fresh start for TMS. Yet quality problems kept popping up: less-than-perfect maneuverability, poor transmission synchronization, insufficient braking durability, and water leaks. Success remained beyond our reach. We had to wait until 1965 to launch our third model. In retrospect, it is clear the underlying cause of the Tiara failure was that Toyota in Japan had rushed its development and had brought it to the market without adequate testing or evaluation.

In May 1965, our third model, a new version of the Corona, was introduced into the American market. We were confident it would outdo other imports in terms of performance and quality. While sales of Toyotas did grow steadily in the U.S. after its introduction, it still fell short of our expectations. This disappointment underscored how difficult it is to overcome the stigma of a bad image once it is established. Only after the Corolla was introduced, as a follow-up to the Corona in April 1968, did we finally embark on the path to success. The first oil crisis in 1973 turned out to be the key, as it boosted demand for a small car with excellent mileage. After fifteen long years, Toyota at last put down roots in the American market.

I returned to Japan in February 1960 and was put in charge of domestic sales. I was later in charge of export and domestic product planning and continued to keep an eye on U.S. market trends and travel there on business. I stayed in the U.S. for four years. Those years were a great experience for me.

I-3. Overtaking the Datsun Bluebird

Upon my return to Japan, I became manager of Toyopet Dealers Section, which supervised the exclusive Corona sales channel. Originally the Toyopet sales channel had been set up in 1956, aiming to ensure high-volume sales for the ToyoAce, our low-priced four-wheel truck. In the previous year, Japan had seen sales of approximately 7,000 three-wheeled trucks per month, while only about 200 SKBs (Toyota's compact four-wheeled truck, renamed the ToyoAce in 1956) were sold. However, mass production permitted a price reduction in January and May 1956 and February 1957. They were sold at under 500,000 yen, a price that could stand up to that of the three-wheeled trucks. Added to this was the sales strength of the new outlets. By April 1957, monthly output had reached 2,000 units. Sales for the entire year exceeded 20,000 units and accounted for approximately one-third of Toyota's total annual sales.

The Corona (a compact car with unitized construction body, smaller than the Crown) was released into the market in July 1957 as a passenger car handled exclusively by the Toyopet dealers. Its role was to take on the Datsun. However, the Corona engine lacked power and maneuverability. It was not highly regarded as a taxi, while Datsun was very popular with the taxi market. The new Datsun Bluebird, an improved version of the Datsun brought out in 1959, opened up a big lead for Nissan. The new Corona came on the market immediately after I was put in charge of Toyopet dealers in April 1960. This car was the domestic version of the Tiara, an export model designed for the U.S. that we had put through full-scale testing at the end of the previous year. The instability of the car body and the poor maneuverability detected during tests were immediately apparent in Japan too, of course. We had to take the time to create a better car.

We used all kinds of technical innovations and sales advantages to try to surpass the Bluebird before we produced a new car. We changed the cantilever-type rear suspension to a leaf-spring type to improve maneuverability, and we bolstered the engine power. Kiyohiko Miura, manager of the Sales Promotion Department (later chairman of Toyota Parts Distributor Tokyo) played stunt man by soaring a Corona over some barrels. This stunt was aired on TV as a commercial to rid the car of its rickety image. In an effort to catch up with rivals prior to the launch of the new model and smooth its path to victory, our sales outlets and staff got together to launch our "Overtake the Datsun Bluebird" campaign. The new model was brought out just as real energy was beginning to build among our sales people.

The new model totally satisfied our expectations and those of the chief engineer, Atsushi Tajima (later managing director of Asian-Warner). Yet we needed to come up with a unique way to show this excellence to potential customers.

By a fortunate coincidence, a portion of Japan's first high-speed road exclusively for motor vehicle use (a 120-kilometer stretch from Nagoya to Nishinomiya) was to be opened in the fall of 1960. This road linked the two economic centers of Nagoya and Osaka. It was the first demonstration of exactly how a speedway could be enjoyed. Even though the new highway was to post a speed limit of 100 km/hr, domestic models could not run at those speeds. Many said that if they could, they would not hold up. To counter these pessimists I came up with the idea of holding an open 100,000-kilometer continual high-speed test on this road for the new Corona (known in-house as the RT40). The general consensus was that a family car should have a life of 100,000 kilometers before it had to be scrapped. Thus, if the Corona could travel this distance without incident, it would go a long way toward building confidence in our product among the public.

When I presented my idea to Toru Inagawa, general manager of the Engineering Department at Toyota Motor (later senior managing director and chairman of the Kanto Automobile Co.), he immediately approved it. I was impressed with the absolute confidence he had in this auto because he was the person with ultimate responsibility for engineering at Toyota. The new Corona model was announced and launched into the market at the beginning of September. Then, starting on September 14, continual day and night runs were begun between Nagoya's Ichinomiya Interchange and Nishinomiya. These runs went on for 58 days and covered a distance of 100,000 kilometers. They clearly demonstrated the high-speed endurance of the new Corona.

The Corona overtook its rival, the Bluebird, that year in monthly sales and became the best-selling car in Japan in 1965. It held on to this position until 1970, when it was overtaken by the Toyota Corolla (a new subcompact).

The battle that raged between the Bluebird and the Corona was dubbed by journalists the "BC War." The ultimate victor was the Corona (Chart 1).

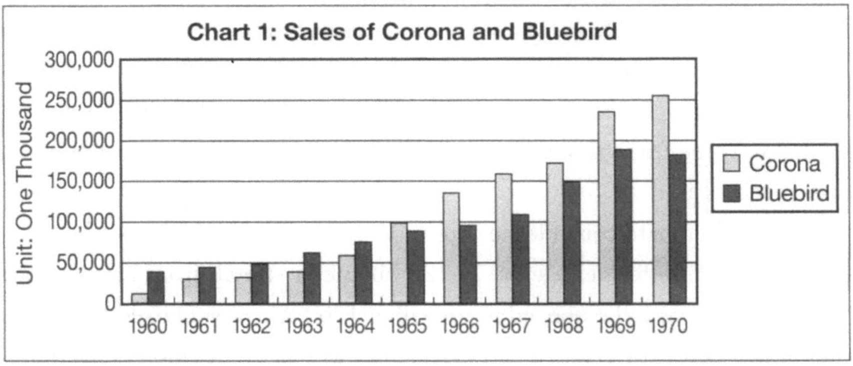

I-4. Failure in Introducing an American-Style Distribution System

Around the time MITI's People's Car Plan was announced, chief engineer Tatsuo Hasegawa (later senior managing director) took over small-car research and development from chief engineer Tozo Yabuta, who had designed Toyota's first prototype small car.

Hasegawa cut down on the weight of the vehicle by changing the traction from front-wheel to rear-wheel drive; he also used three-layered leaf spring in the rear suspension and made substantial use of a light alloy die cast. Comparable foreign models generally weighed about 640 kg.; Hasegawa got the weight down to 580 kg. In June 1960, Toyota put the new car, named Publica, on the market. Though the Publica was lightweight, it didn't scrimp on performance. The retail price was comparatively low at 389,000 yen ($1,072) in Tokyo.

Publica sales were less than satisfactory. A number of mini-cars had been developed at the time, and their presence on the market proved a disadvantage for Publica. Mini-cars had been granted numerous legal incentives such as drivers' licenses, registration, parking, possession tax, and commodity tax. The Publica, closest to these light autos in terms of size and price, was adversely affected by these incentives.

As its prototypes were test-driven more than one million kilometers, the car had proven itself to be tough and durable, with superior fuel consumption. While the public acknowledged the quality, we received many complaints that the car had no aesthetic appeal. In response to these complaints, in July 1963 we put out a completely accessorized deluxe version that drew a sharper line between mini-cars and the Publica. The reaction was immediate, as proven by the 3,700 units sold in December 1963. These figures, coupled with the 3,300 units of the station wagon type (introduced in May 1962), gave us a monthly sales total of more than 7,000 units. We had finally established a market base.

When the Publica was introduced, Toyota set up a sales channel dedicated exclusively to Publica dealers. In the course of establishing this sales channel, we tried to introduce an American-style distribution system. Our plan called for a multiple distributor system. Before then, only one distributor had been established per prefecture per channel, but now we allowed more than one new store in each prefecture, with the area designated as "open territory." This marked the development of an American-style sales network that offered the chance for anyone with the drive to manage his/her own store. Another feature of the system was that sales from the wholesaler to the outlet were conducted on a cash-on-delivery (COD) basis. Before then, the sales outlet would issue a promissory note when the vehicle was shipped from the manufacturer. The payment date was 42 days after receipt of the vehicle, and interest was calculated at 1% above the official rate.

Unfortunately, the introduction of the American-style sales system was a failure in both aspects. The open territory system did not take well. Small sales outlets were established in each region including Tokyo and Osaka, but many ran into difficulty. Weak outlets had to team up with those with a stronger cash flow. Gradually, larger-scale outlets were established while some outlets left the auto sales business entirely. Today, either one outlet handles sales for an entire prefecture or the prefecture is divided into a number of regions, with each outlet having exclusive control over its own region.

The COD practice also did not last for very long. American companies had developed a merchandise collateral financing system in which sales outlets could stock merchandise up to an amount which a finance company or bank approved as a loan. Thus, upon delivery the outlet automatically received a loan which was used to cover the COD payment. However, Japan did not have a merchandise collateral financing system. But stock financing was not a big problem, because manufacturers supplied capital for new car inventory by receiving promissory notes from their outlets. The COD system never took root, so we ended up going back to our original financing method after a year. At present Toyota sells its vehicles at wholesale to all its outlets and receives payment 60 days after shipment.

While the problem with the merchandise financing system was one reason for the failure of the small stores, their failure can be also be attributed to the following problems:

- They did not have enough employees, especially salespeople and mechanics. In the ensuing competition for the available labor pool, large stores clearly had the upper hand.
- A fairly large number of customers purchased vehicles not from the sales outlets allied with the manufacturers but through intermediaries such as neighborhood repair shops. In the Tokyo area more than 30% of all new cars were purchased this way. These intermediaries wanted to deal with sales outlets which offered the best conditions. Under the multiple outlet system, the sales outlets became embroiled in a rebate war, then found it difficult to turn a profit.
- Because of the considerable work involved in financing, taxes, and general business, every Japanese company, regardless of its sales volume, needs a business division. Consequently, the smaller the scale of operations, the greater the burden posed by departments not directly related to sales. The result is an increased share of fixed costs, possibly threatening operations.

And so it was that our attempt to introduce the American sales system into Japan resulted in failure.

Around 1962, Toyota called on its sales outlets to adopt the American add-on rate when calculating interest. The add-on rate makes it easy to calculate nominal interest and creates the impression that the interest rate is low. Immediately this type of calculation spread not only for autos but throughout consumer financing. Consumer groups, though, were complaining that these interest rates were a misrepresentation and called for legal regulations requiring that advertisements clearly indicate that the rate is an add-on rate and show the actual rate. The add-on rate has since dropped out of use because consumers can readily calculate the real interest rate with electronic calculators.

All Toyota's attempts at introducing an American-style sales system thus ended in failure, for we tried to adopt systems without making changes. We did discover from these experiences that Japanese and American business environments are different. As American manufacturers launch a frontal sales assault on Japan, the focus is on what constitutes the most up-to-date American sales system and to what extent it can take root in Japan. Loan conditions today are completely different from when Toyota failed in its attempt to introduce American methods, so it is unlikely that American manufacturers will repeat the same mistakes.

It was only when reading Eiji Toyoda's autobiography, *Toyota: Fifty Years In Motion* (Kodansha International, 1987), that I learned Toyota had proposed creating a joint venture with Ford in 1960 to produce the Publica. Toyota Motors and Ford would both have put up 40%, with Toyota Motor Sales investing the remaining 20%. However, at the end of one year of negotiations the board of directors at Ford rejected the deal, and the joint venture never came to pass. At the time, Toyota, despite its lack of strength, was betting heavily on the Publica. If this deal had been signed, the face of the Japanese and American auto industries would be dramatically different today.

I-5. The Product Planning Office at Toyota Motor Sales: Placing the Customer First

In 1960 I was appointed both manager of the Toyopet section in the Vehicle Department and manager of the newly established Vehicle Body Section. At the time, trucks, with the exception of the standard compact model, were shipped from the manufacturer as a bare chassis or cab-and-chassis. After the sales outlet had received an order from a customer, it would place an order with a body manufacturer to finish the truck. The objective in establishing the Vehicle Body Section was to develop, manufacture, and outfit truck bodies together with truck body manufacturers, then to wholesale them to a sales outlet as completed vehicles. The section itself was tiny (only three regular staff members — two engineers and one female office worker), and it lasted only four years, after which it was converted into the Product Planning Office. Over the four years the engineering staff, which showed remarkable talent, included Yasuhiko Sawairi, (currently managing director of the Automobile Research Center), Tatsuya Takahashi (currently consultant for Ishida Taisei Co.), Shigetada Hanaoka (currently senior managing director of Tokyo Toyota), and Fumihisa Muraji (currently president of Toyota Parts Distributor Ishikawa). All these engineers were involved in the planning, development, testing, and sales promotion of new merchandise. The sole female employee is now the wife of Konen Suzuki, president of Ford Japan, who has put the Japanese auto industry on its ear.

Powerful specialized companies were already manufacturing special-purpose vehicles like fire engines, tank lorries, dump trucks, and concrete mixers. They had established a close relationship with their users and/or sales outlets. This left very little room for Toyota Motor Sales to maneuver. We decided to turn our attention to the development of new products, such as small freezer or refrigerator trucks, or to fields that no one company yet dominated. The manufacturing technology required for the compact van was still in the early stages, so the development process was marked by struggles. But the vehicles that resulted from these excursions into unexplored paths are now manufactured and sold in large volumes.

Road paving was moving ahead smoothly in Japan around this time. But despite all the new paving, roads were not marked with white lines to divide traffic into separate lanes, and it was not uncommon for drivers to drive on either side of the road. Naturally this caused considerable confusion and traffic jams. We developed the road marker to tackle this problem. It came equipped with an arm which extended from the front of the truck; paint was sprayed through the tip of the arm. The truck could lay down a 30-kilometer white line in one hour. While it was highly efficient, the lines it painted were not absolutely straight, so it was not well-liked by the workers. The same type of truck was in widespread use in the United States. Our truck performed about on a par with these, but the degree of precision tolerated in the U.S. was not acceptable in Japan. Despite this, the trucks were purchased and used by many local government construction agencies across the country, even though a number of small manufacturers had already embarked on development of similar devices. A division manager at the Ministry of Construction at one point went so far as to wonder whether Toyota really needed to extend its reach into such a small field.

We also developed an airport limousine, a revamped Crown, which could accommodate more passengers and luggage than an ordinary sedan. We lengthened the Crown 50 centimeters, gave it three side doors on its left side and a rear gate door, and made it roomy enough for three rows of seats and luggage space. This car was shown at the October 1962 Tokyo Motor Show, where it enjoyed attention as the first domestically produced limousine. All the changes in the car body were done by hand. A total of 20 vehicles were manufactured and sold.

The Vehicle Body Section also developed a snowplow that could be attached to the front of a four-wheel-drive Land Cruiser. The same type of vehicle was quite popular in the U.S. It proved to be a success in Hokkaido, which has dry snow similar to that in the U.S., but it was next to useless in the snow belt along the Sea of Japan where the snow is too heavy. Sales outlets ended up trying to figure out what to do with their unsold stock.

A major force behind the development and manufacturing of new prototypes was Sadatoshi Miyajima, senior managing director of the Toyota Service Center (currently known as TechnoCraft, the center was jointly established by Toyota Motor and Toyota Motor Sales). We also received much assistance from companies like Toyoda Automatic Loom Works and Fuji Motors Co.

While we were promoting development of these somewhat minor products, the late Sadazo Yamamoto, managing director (in charge of domestic sales, later president), established an informal sales planning committee in 1962, in order to bridge the marketing strategy of Toyota Motor Sales and the product planning of Toyota Motors. He placed the Vehicle Body Section in charge of the committee's secretariat. Right after the opening of the committee, I presented a new product plan for Corona. The new Corona, which was scheduled to go on the market in 1964, would not be limited to a four-door sedan model. Other versions would include a two-door hardtop and a five-door wagon. This development strategy was meant to diversify the model lineup and enable Toyota to meet a wide range of customer needs. From a designer in the Sales Promotion Department I obtained profile drawings which showed the special features of each model. I took these to president Kamiya and got his approval on the plan. I then went to see vice president Eiji Toyoda with managing director Yamamoto and presented him with our ideas. Toyoda gave his assent, remarking that it was about the right time for Toyota to produce just these types of car.

The new Corona sedan was put on the market in the fall of 1964. In July 1965, it was followed by Japan's first hardtop, then the wagon in October. These models made the difference between the Corona and the Bluebird all the more clear.

In 1964, the informal sales planning committee was made into the more official Product Planning Office of the Vehicle Department. The Vehicle Body Section was disbanded and incorporated into the new office. I was given responsibility for running this unit. In 1979 the office was moved out of the Vehicle Department and put directly under the president. During this period I was promoted to director of the board and later managing director. I remained linked to the management of the Product Planning Office from beginning to end. Consequently I was involved in Toyota product development for approximately twenty years, from the latter half of the 1960s through the first half of the 1980s.

It was difficult to convey to the rank and file throughout the company the importance of product planning, but I had the complete cooperation of the Personnel Department, which passed on to us top-notch personnel from among incoming employees and the entire employee pool. These people had a passion for cars and a talent for planning. Tatsuro Toyoda (former president, TMC), Terukazu Inoue (standing auditor, TMC), Tatsuro Iwakiri (president, Miyazaki Transport Co.), Mitsutoshi Otobe (managing director, Nagoya Broadcasting Networld), Hiroaki Kunita (general manager, Product Management Division, TMC), Takeshi Shiino (former president, Toyota Kanto Automobile Vocational School), Masamichi Susuki (senior managing director, Toyota Parts Distributor Gunma), Hiromitsu Matsumoto (director and president, Kansai Automobile Vocational School), Yoshiharu Matsumoto (standing auditor, Institute for International Economic Studies), Koji Mitani (director, TMC), Tomoyuki Mizuno (deputy general manager, Toyota Overseas C&A Division,

TMC), Hideaki Miyahara (director, TMC) and Mohei Harada (executive vice chairman, Fuji Auto Body) all spent time in this office. All new members were greeted with the admonition, "You are not to talk even with family members or friends within the company about what goes on here. If you don't think you can adhere to this rule, you're not welcome here." In contrast, all information within the office flowed freely, and members were encouraged to offer their opinions even on vehicles they were not directly responsible for. In this way we obtained valuable insights from many people while also getting a fresh perspective from the younger staff members.

I also insisted that all staff members become well acquainted with everything going on in Detroit. We had them monitor the goings-on in the American auto industry by obtaining and studying information from newspapers and magazines. We used computers to collate this information and then used the data in developing our sales strategy. We sent some of the newcomers to the Computer Department, where they received training in computers for one to two years. These trainees then drew up various simulation models for us, meaning we did not have to rely on Computer Department staff. When their computers were available to us, our trainees could operate the machines and do valuable work for us. Some of the following are examples of simulation models we came up with. Section manager Terukazu Inoue and Hideaki Miyahara teamed up to try to run through a war game which had been developed by the U.S. Department of Defense. Professor Yutaka Osawa of Osaka University gave them good advice and helped them to finish the job. They created a large-scale simulation model while analyzing and researching Nissan's corporate activities, as a way to better understand the competitive strengths of Toyota and Nissan. The IBM 7040 was the only machine at the time which could handle such a large-scale model, and the only way to use it was to rent time with IBM Japan in Tokyo. The endeavor took an incredible amount of effort. A single mistake in the system design or a mispunched card meant that the computer wouldn't run. Thus, a countless number of trips had to be made between Nagoya, where we were located, and Tokyo. These efforts were worthwhile, however, as the simulation showed that our superiority over Nissan would not be jeopardized.

I also conducted regular training exercises, such as creating logistics models, so that the computer could be put to use in drafting sales strategies.

I sent our younger staff members to sales outlets and customers in Japan and overseas as well, to give them market experience. I told them not only to hear what distributors and dealers wanted, but also to get behind the wheel of our cars and our competitors'. I expected them to make their own market evaluations on the products and the sales regions. In this way they also gained intimate knowledge of overseas roads and traffic conditions. Some employees resided in the same place for about six months. By receiving Toyota Motors chief engineers and technicians overseas, and allowing them to become familiar with key personnel and points, they ended up understanding the needs of the market in a short period of time.

Through this training I had the best marketing staff in the world. They knew not only the Japanese market inside out, but overseas markets as well (including, of course, the U.S.).

I would like to present here the approach of the Product Planning Office, which I will illustrate in greater detail later on. The better the engineer, the stronger his or her point of view and, unfortunately, the less likely he or she is to seriously consider the opinions and advice of others. Consequently, I drilled it into our staff that, for example, even if they had some good advice to give, it was best to convey this advice by interacting with engineers and letting them think it was their own idea. The Product Planning Office was to remain in the background. This was brought up by several of my subordinates during a gathering marking the 30th anniversary of the founding of the Product Planning Office.

During the course of a discussion within the Product Planning Office, I would never immediately say yes to any new proposal. Even if I thought it was a good plan, I would first bombard the proponent with questions. If the plan was truly good, its backer would hold firm, and in the process of defending it he would hone it into something better and fresher. I would wait until that point before giving my approval.

Once a Product Planning Office plan was drawn up, Engineering Department factory staff would be contacted unofficially through the Domestic Sales Administration Department, which served as Toyota Motors' liaison for

Toyota Motor Sales. In this way the chief engineer in charge would incorporate our plan into their own plan. Finally it would be presented at the monthly meeting of the Joint Toyota Motors-Toyota Motor Sales New Product Committee, which was attended by top management from both companies. The plan would be reviewed and decided upon. Meanwhile, the Product Planning Committee within Toyota Motor Sales, made up of officers at the managing-director level and above, would meet every month to hear reports on plans on which we had agreed with the design and engineering staff of Toyota Motors. This was a good opportunity for top people preparing for the joint committee meeting. Products were developed following these procedures. However, since the Product Planning Office rarely stepped into the forefront of the process, very few executives in either company truly understood the details of the work in process.

The Product Planning Office was able to function effectively under a system in which Toyota Motors and Toyota Motor Sales operated independently because of the respect that the top executives in Toyota Motors had for president Kamiya. One of a chief engineer's most important tasks was to get the approval of president Kamiya directly or through then vice president Eiji Toyoda when a new car plan was proposed within Toyota Motors. The Product Planning Office served as a bridge between the chief engineer and president Kamiya. The president did not allow executives in charge of sales operations to state their opinions directly to a chief engineer in Toyota Motor. The opinions of Toyota Motor Sales had to be conveyed through the Product Planning Office.

It was through this approach that the Product Planning Office was able to fully execute its tasks in a relationship coexisting with both Toyota Motors and Toyota Motor Sales.

I-6. Corolla vs. Sunny

As I have mentioned, almost as soon as the Publica was put on the market we received complaints from consumers and front-line dealerships that its interior was narrow and clumsy. They wanted a more spacious car. I had the Product Planning Office draw up a diagram showing the sales results for various vehicles in different price ranges, looking at both new and used cars. The resulting diagram revealed a considerable concentration of vehicles in the layer above the Publica and below the Corona. We determined that the largest consumer market was located in the price layer above the Publica. We immediately embarked on the development of a new auto — later called the Corolla — for the general public, aiming at this market. This new project was also pushed forward by chief engineer Hasegawa, who had been in charge of the Publica.

The first major consideration was how large the auto should be. The Publica's interior was definitely cramped and clearly tested the patience of passengers. Thus, the problem became just how much the car should be enlarged. A survey of Publica owners in Tokyo showed that, almost without exception, their garages could accommodate a midsized vehicle like the Crown. This information indicated that the main reason behind their purchase of the Publica was probably not its size but its low price. Still, they could afford to pay a bit more for a roomier car.

Taking into consideration the U.S. market, chief engineer Hasegawa proposed that the Corolla be designed with taller Americans in mind. He wanted the front seats to be able to comfortably accommodate a tall individual, though the back seats could be slightly cramped. However, president Kamiya felt that such a car might too big. He felt that if the vehicle was too large, it would add to the cost and reduce the potential purchasing pool. He wanted any increase in size to be kept to an absolute minimum.

The Product Planning Office went out and bought some imports in this class and checked their external and internal specifications. The results of the investigation tended to support Hasegawa's position. Hasegawa's proposed vehicle was smaller than the British Vauxhall Viva, but the dimensions for the front half were about the same. We then taped a silhouette of Hasegawa's car onto a Vauxhall silhouette and showed it to president Kamiya. He gave his approval after climbing into both the front and rear seats.

The executives at Toyota Motors and Toyota Motor Sales also asked us to ensure a clear distinction between the Corolla and the Publica in terms of class and price. Even after the introduction of the Corolla, they wanted not only to maintain but also to expand the sales of Publica, which represented much hard work and effort. We also conducted research together with Toyota Motor and Toyota Motor Sales to establish a difference in price that would allow both cars to coexist. This research was conducted under the direction of Professor Jiro Kondo of the University of Tokyo (former chairman of the Japan Academy of Science) based on a method called Operation Research (OR) which had been developed during the war by the American military.

While we in the Product Planning Office were moving forward with the development of the vehicle, Toyota embarked on the construction of a plant in Takaoka in December 1965. This plant was to be dedicated exclusively to the Corolla and have a monthly production capacity of 16,000 units. At almost the same time, the Kamigo Plant was built to produce the Corolla engine. In the end, the Corolla achieved average monthly sales of 20,000 units. Looking back, I am amazed that Toyota was willing to set forth on such a venture at a time when monthly sales of the Publica had only reached 4,000 units.

I ran into Eiji Toyoda, honorary chairman of Toyota, at a party in January 1994 and asked him what made this decision possible. He responded that he gave the go-ahead after talking with Kamiya because the two of them agreed that a car that targeted the average consumer would sell well. It was then that I came to understand Kamiya's great drive and tenacity when it came to the Corolla. His mission was to capture the sales necessary to get Corolla plants working at 100% capacity. He was relentless in ensuring the unequivocal success of the Corolla in both production and sales.

We also found out that Nissan at this time had abandoned its belief that a car for the general public was not needed and was moving forward in developing the Sunny, a vehicle similar to the Corolla, aimed at the average consumer. The vehicle ownership tax in Japan then was based on the size of the engine — the bigger the engine, the bigger the tax. The lowest ownership tax was on engines with a cylinder capacity of 1000 cc or less. Thus, everyone involved in the development of a low-priced vehicle for the mass market held the unquestioned belief that the engine should be 1000 cc or less to take advantage of the lowest ownership tax rate. The design of the Corolla engine was also based on this assumption.

Nissan thought their 1000-cc class car aimed at the mass market was an assault on the Publica's dominance. Yet the difference between the tax on the 1000-cc or lower engine and those engines which went up to 1500 cc was only 1,500 yen per year. Thus we determined that the ownership tax did not present a real obstacle to building an engine with the optimal capacity, despite all the talk of a 1000-cc barrier.

It was easy to expand the already developed 1000-cc engine to 1100-cc without changing the basic design. Even taking into consideration the production facilities, we ascertained that it would be relatively simple to increase engine capacity. As soon as we confirmed that the Sunny, scheduled to be released in the spring of 1966, would have a 1000-cc engine, I got together with chief engineer Hasegawa in the Engineering Department and discussed changing the Corolla to an 1100-cc engine, even though it was only six months before we were to enter full-scale production.

He asked me to get the approval of president Kamiya, so I jumped on the bullet train to Tokyo and presented him the revised plan. He immediately gave it his approval. I then jumped back on the bullet train, returned to Nagoya, and late that evening told Hasegawa that we had the go-ahead. However, when I talked to Hasegawa about this much later, after he had left the position of chief engineer, he remembered a different version. He recalled that the plan had been settled upon during discussions with Eiji Toyoda, then vice president. If this was the case, then it would seem that by the time I talked to Hasegawa the decision had already been made at Toyota Motors, and the only thing that must have remained was the approval of president Kamiya.

Although Kamiya had given his approval, he believed that even a slightly lower price was critical to high sales. Until the very end he seemed unsure as to whether the idea to increase engine size was a good one. I later learned from the president's secretary that his yes can sometimes mean no, and perhaps that was true in this case.

As soon as the first stage of work was completed at the Takaoka Plant in September, 1966, we went into production. The first Corollas were sold in October of the same year.

The Sales Promotion Department, led by Mikio Hayashi (later managing director and chairman of Toyota Used Car Sales Co.), came up with the catchy slogan, "An Extra 100 cc Allowance," and started to take on the Sunny, which had about a six-month head start. Monthly sales of the Corolla exceeded 10,000 units by July 1967, and annual sales outstripped those of the Sunny. By 1968, the Corolla had overtaken the Bluebird. Annual sales exceeded 240,000 units in 1970, making the Corolla the best-selling mass-produced passenger car in Japan, a position it has maintained ever since (Chart 2).

Exports of the new Corolla to the United States began in April 1968. The export of Toyotas that year, including our new car, exceeded 90,000 units. Toyota had finally put down roots in the U.S. market. The Corolla was also exported to regions throughout Europe and Asia, and became known as a "global car." It overtook the Volkswagen in 1974 and has since been the best-selling car in the world.

Though the Publica ran into a variety of difficulties after the introduction of the Corolla, it remained firmly entrenched as the first Toyota a driver was likely to buy.

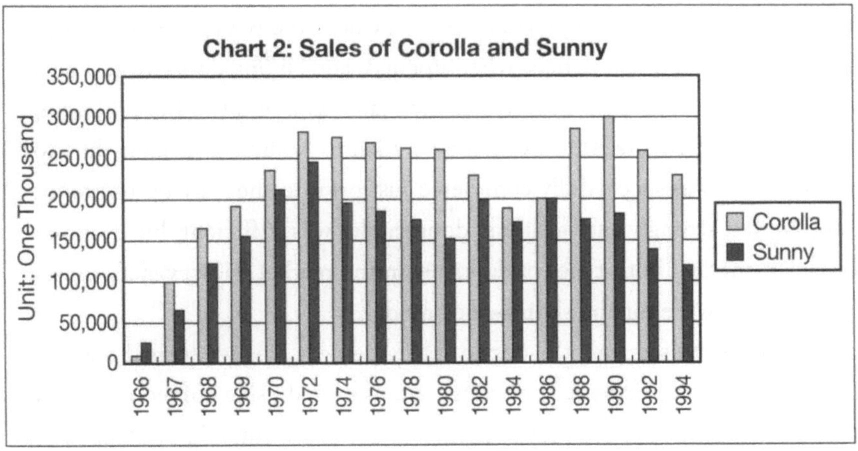

I-7. Foresight Behind the Development of the One-Box Van

Right after the Automotive Sales Product Planning Office was established, I was taken to one part of the Toyota Auto Body's Kariya Plant by chief engineer Saihei Iritani of Toyota Motor (later managing director). A single compact van was parked in this cramped, gloomy, warehouse-like plant. Both sides of the vehicle were flat, it had a high height-to-width ratio, and it didn't appear very sturdy. The prototype had been hammered out by hand, but the finishing work was actually quite good. It was based on a frameless unit construction with a wishbone independent front suspension with coil springs and a torsion bar stabilizer. Iritani asked me if I thought that monthly sales of 2,000 units were possible.

While this type of one-box compact van was quite popular in Europe, made by everyone from Volkswagen to Benz, Japan represented virgin territory. I thought that the 2,000 figure would be difficult to achieve, but for those of us on the sales side, nothing gave us greater pleasure than the birth of a new type of vehicle. I promised to look into the matter and see if there was a way to make their figure come true.

The first thing that the Product Planning Office considered was how to boost the commercial appeal of the vehicle. We started by looking at whether we could install sliding doors. The few Volkswagen vans imported into Japan had doors that opened on hinges. Under the wrong conditions these doors could pose a real danger when opened, and sliding doors would be a great sales point in Japan, with its narrow roads. As a result of the unflagging efforts made at Toyota Auto Body, led by Kichiro Kojima, managing director, development was successfully completed just prior to the start-up of production. We felt strongly that a sliding door model was sufficient, but Kojima was adamant that both types of doors be on the market. However, all subsequent orders were for the sliding-door model.

Prior to this, during the design stage, the next area we turned our attention to was how to upgrade the styling. We proposed using curved glass in the side windows, but were not given permission because curved glass had not yet been used in passenger cars. As soon as this van was put on the market, Volkswagen came out with a model equipped with curved windows. This proved that our request was in tune with the times, but all we could do was wait for the next model to be released.

Our third area of concern was to provide as much variety as possible. We agreed to develop three-seat and six-seat delivery vans, a nine-seat wagon, and a 12-seat and 15-seat compact bus (dubbed the Commuter). Nevertheless, it was difficult to sell 2,000 units a month, and the vans were produced together with one-ton frame trucks. Their size overlapped with the ToyoAce, but their low deck and pleasant interior were their main sales points.

In October 1967, the one-box van was put on the market under the name HiAce, with a sales target of 1,200 van-type vehicles and 800 trucks. Sales were brisk in the final three months of the year, totaling 7,000 and then exceeding 40,000 in 1968. The curtain was up on the era of the one-box van.

Looking back on it today, Toyota showed astonishing foresight in developing the unit construction one-box van around the time when a box-shaped van, mounted on the rear of a truck with a small cabin, had been introduced on the market. Recently I discovered it was then vice president Eiji Toyoda who came up with the idea for this vehicle. Chief engineer Iritani and Isamu Shibata (later a chief engineer working in compact trucks, then president of Takashimaya Nippatsu Co.) received detailed instructions from vice president Toyoda, who encouraged them to "give it a try — I am certain the two of you can do it." When they were given the go-ahead, the prototype had already been completed. It had secretly been assembled at Toyota Auto Body in accordance with orders issued by Toyoda to chief engineer Tozo Yabuta. Toyoda first ensured that it was technically possible himself, and then officially issued orders for the car to be put into production.

I believe that this was the first and last time this degree of variety was made available from the moment a vehicle hit the market. I am still extremely grateful to all those involved at Toyota Motors and Toyota Auto Body. They listened seriously to what we wanted for this vehicle and cooperated in its development. I recently happened to meet chief engineer Isamu Shibata at a gathering of "Friends of Toyota," and we reflected back on those days. We reminisced about how obsessed we were with this project and how late we worked almost every night without complaint. The Product Planning Office demanded quite a lot from the engineers, who successfully achieved the goals we set for them. I take pride in the fact that the original models are still around, 30 years later. This shows we worked in the right direction.

This success gave us the courage to turn our attention from the MiniAce to the LiteAce. The MiniAce was a small truck with a horizontally opposed two-cylinder engine (used in the Publica) mounted under the floor. At this time we were under assault from mini-vehicles, which were granted legal incentives as mentioned earlier. Sales were unsteady, and outlets were asking for a far-reaching counterattack. Our response was the development of the LiteAce, a commercial van aimed at the average consumer. Toyota Auto Body was in charge of development, and its managing director, Mitsuo Nishida, visited the Product Planning Office himself innumerable times with the drawings tucked under his arm. He gave time and careful consideration to what we wanted in the vehicle as the design and construction of the prototype moved ahead. Production was carried out at Daihatsu, and the van was put on the market in November 1970. A total of 27,000 units were sold in 1971 and more than 40,000 in 1972.

The advent of the light van boom was clear to all. Visitors from foreign auto companies were astonished by the stunning array of light vans on the streets. Tokyo had become a city of vans. It got to the point where each of the four sales channels was demanding its own van model to sell. The question became whether there was a separate market for the HiAce and the LiteAce. We conducted an extensive analysis and vigorously debated the issue in the Product Planning Office. Our eventual conclusion was that there was, in terms of both size and price.

We determined that even with the addition of another van oriented to this market, three light van models differing in size and price could coexist. So the current TownAce was proposed and developed by Toyota Motors. The vehicle was introduced in October 1976, and more than 50,000 units were sold the following year. Sales of the TownAce grew steadily with barely any effect on the HiAce and LiteAce. Over the next 10 years Toyota created a one-box light van market that grew to annual sales of more than 150,000 units. Of course, this substantially strengthened Toyota's domestic sales. None of our competitors launched a serious challenge in this area, because they either had not noticed the market or didn't have the drive to embark on development.

The one-box, refitted to accommodate passengers, had begun to show real signs of popularity even before the bulging of the bubble economy. It was normal for customers to have to wait three months for their orders to be filled. In some cases, they had to wait double that. Looking back on it today, this was a presage of the current recreational vehicle (RV) boom. I regret that we didn't exploit it sooner. We gave up the lead to other companies in the development of a sport utility vehicle (SUV), although the RAV4, which we put on the market in 1994, has been doing well. I look forward to our efforts in the future.

Six years after the launch of the HiAce, Toyota introduced a bonnet pickup truck, the HiLux. This market had been the exclusive domain of Nissan. Their Datsun trucks had a legion of loyal customers. We knew that to prevail over them would not be an easy task. It took 20 years of stubborn effort to catch up and overtake them (we did so in the latter half of the 1980s). The introduction of four-wheel-drive was a big factor in this victory. Toyota and Nissan have both succeeded in producing and selling these pickup trucks on-site in the United States.

I-8. The Shock of "My Years with General Motors"

I think it was in the spring of 1964 that I came across an article introducing the book *My Years with General Motors* by the former chairman of General Motors, Alfred Sloan, Jr. (Doubleday, 1963). The article was in a journal put out by the National Automobile Dealer's Association (NADA). I immediately had the book sent over by Toyota Motor Sales, U.S.A. This book revealed to me exactly what Toyota had to do at exactly what time. Toyota had achieved its goal of putting together a lineup of passenger cars that covered a spectrum of customers: first buyers, those looking for a mid-range car, and others wanting luxury—namely the Publica, Corolla, Corona, and Crown. The book came out just when we were searching for what to tackle next.

Sloan had spent considerable effort creating an automotive giant through mergers and acquisitions, linking entities together through a management system. Since we at Toyota also aspired to grow, it occurred to me that if we took to heart the struggle, experience, and advice contained in this book, we could have a smoother ride to success than GM had had. For example, if we worked hard developing new cars and new sales networks, giving careful consideration to priorities, we could expand our reach in a systematic fashion. Continually touching base with Sloan's finely honed strategies, we were able to move forward from a vantage point not available to Sloan. It was almost as if the book was an instruction manual written just for us.

The longer our rivals were unaware of this book, of course, the better for us. Fortunately, the Japanese version was not published until October 1967, well after the original came out (1963). Most Japanese people, including managers in the auto industry, were totally ignorant of the existence of this book until it came out in Japanese. The almost-four-year period between its publication in English and its publication in Japanese coincided with a period during which important strategy was formed that decided the subsequent direction of the Japanese auto industry. The fact that we got our hands on this book at an early juncture put us at a considerable advantage.

We took in much of what Sloan had to say, but I would like to quote one passage of particular importance:

> *The product policy we proposed is the one for which General Motors has now long been known. We said first that the corporation should produce a line of cars in each price area, from the lowest price up to one for a strictly high-grade quantity-production car, but we would not get into the fancy-price field with small production; second, that the price steps should not be such as to leave wide gaps in the line, and yet should be great enough to keep their number within reason, so that the greatest advantage of quantity production could be secured; and third, that there should be no duplication by the corporation in the price fields or steps. (p. 65)*

I took a personal interest in this book. Sloan repeatedly emphasized that management approaches must change as the world changes. However, what was of particular interest was the way individuals who inherited great wealth adapted to change in order to keep their wealth intact.

The book was written literally, without allusion to emotion, and might be interpreted to suggest that top executives should be cold-blooded. Professor Peter F. Drucker, a friend of Sloan's, has stated that Sloan is a man full of feelings who continually looks out for his subordinates. However, one would not get that impression from reading his book. It seems his successors at GM have only taken to heart the man reflected in the pages and have forgotten his benevolent side.

As I stated in the preface to this book, I began following GM's movements with great interest for professional and personal reasons. In early summer of 1981, a group of analysts from New York, led by Maryann Keller, came on a fact-finding mission to Japan and visited Toyota. I greeted this group in my role as managing director in charge of public relations. After giving my standard presentation on Japanese automobiles, I ended with my best wishes for the success of GM's new J-Car, dubbed the "Import Fighter," to be released in the next couple of days. Keller asked me how I knew about such goings-on in the United States. My response was that if I didn't know something on that level, I could hardly be in charge of public relations at Toyota.

I-9. The Revolutionary Celica Ordering System

The development of our basic car model lines had produced the Crown, Corona, Corolla, and Publica. However, our inability to come up with a car that was just right for the youth market was becoming a real problem. We had paid close attention to the major success that Ford had achieved in launching the Mustang. The development and sales of this car was promoted by Lee Iacocca, former chairman of Chrysler, during his tenure as general manager in the Ford Division of Ford Motor Company.

The Mustang was a revised version of the Fairlane, a basic model with a European sportiness not previously seen in American cars. The sleekness of a sports car came with a wide range of options and a stunningly low standard price of $2,400. Sales of the Mustang exploded, particularly among younger buyers.

The Mustang brought on a change in passenger car classifications in the U.S. Now, in addition to the conventional full-sized, intermediate, and compact classes, there was the specialty car.

While taking note of the success of the Mustang and the skyrocketing growth of the specialty car market, we ascertained that a similar market could be created in Japan. At the beginning of 1967, we embarked on the creation of the native-born specialty car. This was the Celica. It was produced at the newly constructed Tsutsumi Plant, Toyota's third assembly plant (after the Motomachi and Takaoka Plants), dedicated exclusively to the production of passenger cars. The Celica was introduced into the market in September 1970.

An important matter was how to create a compact specialty car with a distinctive style. The designer of the Celica was lucky—he had to present his plan only once to executives of Toyota Motor and Toyota Motor Sales before getting the OK. Typically, changes are suggested and a design has to be examined a number of times before it is approved. In all my 20 years at Toyota, the Celica is the only vehicle whose design was approved in its first go-round.

The car had originality combined with a freshness and appeal not limited to the youth market. It shared a platform with the Carina sedan which was launched at the same time. It embodied the real flair of a specialty car.

The "full choice" system was also developed in the production and sales of the Celica. This system was a revolutionary concept designed to meet the demands of "the age of individuality." Some American cars came equipped with a broad range of options (dubbed "wide selection"). The Mustang went beyond the level of options offered by such cars, and this served to emphasize its status as a specialty car. However, even though the Mustang manufacturer listed a number of options from which to choose, it did not have a system enabling the production and delivery of a desired product to a customer in a short period of time. While studying how to offer multiple options to underline Celica's status as a specialty car, we hit upon giving the customer a "wide selection" which would, as the phrase suggests, leave the choice to the customer. I discussed the matter with chief engineer Katsuhiko Nishida (currently consultant with Tsuchiya Co.), and he took on the job without hesitation.

The Celica full choice system (a term president Eiji Toyoda coined) offered a full range of options, at specifications to be decided upon by the customer, while ensuring quick production and delivery. The car came with four engine options: SOHC 1400-cc, SOHC 1600-cc, SOHC 1600-cc with twin carburetor, and DOHC. In addition, there were three transmission options (four-speed manual, five-speed manual, and three-speed automatic), four exterior options ("extra touring," "luxury touring," "sports touring," and "ground touring), and six interior option packages. In addition, the customer could select a rear window defroster, leather roof, colored bumpers, radio, heater, clock, curved mirror, stereo, air conditioner, and power windows. Of course, customers could also choose from a wide range of colors and put together a combination that matched their own desires and budget.

We were able to satisfy the desires of the customer through the introduction of the dealer order entry system. We set up an on-line computer link between Toyota Motor and Toyota Motor Sales and between some of our sales outlets and Toyota Motor Sales. The production plan at the factory was adjusted daily in response to customer orders, with the objective of producing a car with a minimum lead time, ideally within 10 days.

We considered computerizing the process to make selection as easy as possible. Professor Yutaka Osawa of Osaka University told us it was possible and showed us how to develop software in the Product Planning Office. Hideaki Miyahara, a member of the office, studied how to use a new IBM computer and developed a program in only six months. We named the program the "Car-puter System" and tested it in sales outlets in Nagoya.

An on-line system linked a mainframe IBM 3600-50 in Toyota headquarters with the Car-puters that were set up in showrooms of sales outlets in Nagoya. The Car-puter was made up of an easy-to-use IBM 2760 and an IBM typewriter. The IBM 2760 was the most up-to-date machine at the time and had been introduced at the Osaka Exposition in 1970. Cars were displayed on a monitor controlled by the computer. The system operated in the following manner:

The customer answered approximately 15 questions covering age, family situation, history of car use, ownership conditions, and driving history. These answers were analyzed by the IBM 2760. It then output what it thought was an appropriate auto. The customer then input his or her budget. The IBM 2740 typed out the optimal car and its scheduled delivery date. If the customer wanted to change the type of car or delivery date, the IBM 2760 output a different selection. The computer, of course, stored the previous selections and made the necessary corrections.

Color slides were displayed on the monitor throughout the above process. Customers were able to easily select a car that met their tastes and budget. Once the customer placed an order, the sales outlet could immediately send that order to the manufacturer through the dealer order entry system. Preparations for manufacturing could be implemented immediately.

The computers of the 1970s were truly primitive; they could hardly be compared to those of today in terms of performance. Images were projected using color slides, as opposed to modern-day CD-ROMs. Toyota was conducting such field tests in 1970, a quarter of a century prior to the advent of the multimedia age.

In the summer of 1971, Elizabeth Sampson, a Columbia University student, was studying at the Product Planning Office as part of the AIESEC (Association of International Exchange Students of Economics and Commerce) student exchange program. The office hosted one student a year from countries such as the United States, Germany, Great Britain, and France. I had these students write reports on the Japanese automobile industry during their two-month stay. All of them attacked their task with vigor and wrote excellent reports. The younger members of the office staff helped them write their reports and took them out on drives and to parties on the weekends. My objective was to give our staff a chance to get used to interacting with foreigners and, of their own volition, to go overseas. Some of these exchange students went on to get their Ph.Ds in economics or entered business or the academic world. Some returned later with their spouses to visit Japan. I kept in contact with them during business trips, even if only for a visit of a few hours.

When Sampson heard about our Celica full choice system, she remarked that she would really like to buy a car using the system and take it back home with her to the 'States. How great it would be, she said, to have the one car in the world that was made just for her. This system was used only in Japan for the Celica, and since the car didn't meet American safety and emission standards (nor come with left-hand steering), we unfortunately had to disappoint her.

In 1972 I heard from Shozo Harada (a university classmate, currently a member of the Lower House), who had attended the Tanaka-Nixon talks in Hawaii, that the U.S. was promoting sales of large airplanes for Lockheed. When I was in Washington attending Transport '70, an international conference covering future technology for airplanes and autos held by the U.S. Department of Transportation, I had been given a tour of a huge transport plane on display at an air show held at the same time. I had been truly impressed by the size. After hearing Harada's story, it struck me that we could use one of these planes to ship Celicas manufactured under the full choice system. We could ship 120 Celicas directly to the U.S. from the factory in such a transport plane. Although delivery would take a week longer than in Japan, we could still provide our American customers with custom-made autos. I thought we had a reasonable chance of success, based at least on the reaction of one young American lady.

I had the Product Planning Office staff look into the matter. They found out that the runway at Nagoya's airport, our point of departure, was a few hundred meters too short for these transport planes to take off. TMS also showed interest in the idea, but it was just at this point that Prime Minister Tanaka was forced to resign as the Lockheed scandal came to light, and the plan lost its backing.

Our full choice system was made possible because production was based on the Toyota production system (known overseas, including in the U.S., as the lean production system). Our production system delivered the necessary parts and components to the assembly line "Just-In-Time," in accordance with the *kanban* method. "Just-In-Time" ensures that subassembly stock is kept to an absolute minimum. In the *kanban* system, required parts are sent at the required time in the required amounts.

Despite the fact that the customer could select from a large number of options, thereby creating practically an infinite number of combinations, each option was broken down into parts and components, and the order volume for each part was quite stable. Thus it was possible to provide a steady flow of option parts, and the production of different cars could be evenly spread over the assembly lines. I think this demonstrates the presence of an "invisible hand" in the market economy that controls economic behavior.

The Celica hit sales of approximately 100,000 units in its first 20 months on the market. We had achieved our original aim of creating a specialty car market. Beginning in 1972, the Celica was exported to the United States, where it was a huge success. It was particularly popular among women, who purchased more than half the units sold.

The Celica represented a new test for Toyota. Its success attests to the organizational and functional flexibility inherent in Toyota at the time and the energy with which we took on new challenges. The full choice system was dropped years later, because the number of models was cut down to cope with a labor shortage which arose as production expanded. However, the Toyota production method has spread rapidly among American and European manufacturers, and the foundation was established for the implementation of such a system among foreign manufacturers. The biggest issue among manufacturers around the world is how to improve customer satisfaction. I think in the future we can expect to see this type of ordering system brought back on-line and made a standard operating procedure.

I-10. Development of Distribution Systems

Starting in the 1960s, industries and corporations began to call for distribution streamlining. As we entered the 1970s, the Product Planning Office received two requests to develop a distribution system. One was from a milk company that wanted us to come up with a way to get milk from their plants to depots in the center of cities, then transport the milk to retailers. The other request was from the Tokyo Metropolitan Department of Sanitation, to come up with a system for transporting garbage from collection points to incinerators.

For the request from the milk company, we developed high-performance two-ton cooler containers. Five of these containers were loaded onto a trailer for shipment from the plant to the depot. Individual cooler containers were then put onto two-ton trucks for transport from the depots to the retailers. The coolers took the place of a warehouse. All that was required was simply paving the yard. There was no need for the usual buildings, new cooling warehouse facilities, or maintenance fees. The stacking and changing of containers could be done quickly with a forklift, and the driver work and wait between loads were both eliminated. The turnaround rate was boosted, and the number of vehicles needed was reduced. The high-performance cooler containers used 50 kilos less ice per day than before. Distribution costs were cut by approximately 15% because no driver's helper was needed, the number of trucks was reduced, and a refrigerated warehouse was no longer needed.

The second system, designed for the Sanitation Department of the Tokyo Metropolitan Government, was the reverse of the first. Our design allowed containers to be placed on the collection vehicles, the garbage was collected from the streets, three of the containers were put onto a large truck at the depot, these containers were transported to an incinerator, and the garbage was dumped with the containers still loaded on the truck. Requirements for the system included no leakage of dirty water or release of foul odors during the collection and transport of the garbage. These requirements made it necessary to come up with some original approaches, and we received 19 different patents for different parts of the system. We were successful in containing contaminated water, foul odors, and garbage, and in speeding up transport.

During the development process for these two distribution systems, the patented truck technology we had developed with related manufacturers came in quite handy. Also, by upgrading this equipment technology, it became clear that Toyota could take on a wider range of roles. At the same time, the development of a software system was essential in facilitating the measurement and evaluation of equipment. In 1971, a Toyota Dynamics Logistics (TDL) team was assembled, centered in the Product Planning Office of Toyota Motor Sales. Other participants included managers from Toyota Motors, Toyoda Automatic Loom Works, Nippon Denso, and Toyota Auto Body. This team focused on both equipment and design-related issues in the development of these systems, working under the leadership of Tomoyuki Mizuno in the Product Planning Office.

The equipment system was called the "Toyota Easy Carry System" (TECS). It was composed of four basic hardware systems, namely a container system, pallet system, chilling system, and specially equipped trucks. These systems were all developed with the help and cooperation of other manufacturers. The software and design aspects were handled mainly by the Product Planning Office under the guidance of Professor Yutaka Osawa of Osaka University. The development of the equipment and hardware aspects were conducted under the direction of Professor Yasushi Tanazawa of Tohoku University and Professor Naomichi Hirayama of Tokyo Metropolitan University, while software development and design were conducted under the direction of Associate Professor Kuniaki Miyamoto of Osaka University (currently dean of Takaoka Junior College). Evaluations were made of potential distribution systems. Plans were drafted covering the number of vehicles, containers, forklifts, and workers required, along with the size of the depot site; then target cost reductions were presented.

Chapter II

A Flagship Car

A Flagship Car

II-1. MITI's Shadowboxing

MITI felt that, as Japan opened up its economy, it was imperative to build a new economic structure that was internationally competitive. The open economy would be achieved through three liberalizations: trade, foreign exchange rates, and capital transactions. In October 1961, the Ministry set up the Industrial Structure Study Council to study the ideal industrial system. (The name of the council was changed to the Industrial Structure Council in 1964.)

To facilitate study of automotive issues, the Passenger Car Subcommittee was established in April 1962. The Japan Automobile Dealers Association, as a member of this subcommittee, took up the issue of sales distribution and chose me to serve on the subcommittee. The motor vehicle had rapidly become a popular form of transport, more quickly than legislation and studies could keep up. I listed several problem areas that should be considered to improve auto sales and user conditions. For example, a comparative study of Japanese and California auto registration systems illustrated that the Japanese system was unnecessarily complicated and that a substantial amount of streamlining was possible. The study also showed that driver's license tests were much more demanding in Japan than in California and that obtaining a license itself was far more involved than necessary. I emphasized that the confusion surrounding legal standards for motor vehicles created obstacles in the design, production, distribution, and export. The subcommittee listened positively to these comments and included them in its interim report, presented on December 13, 1962. My main points were also included in the subcommittee's final report. Yet, although the driver's license testing system was slightly revised in line with the proposals presented, very little was changed.

In October 1962, MITI established the Passenger Car Policy Special Subcommittee within the Industrial Structure Study Council, separating its functions from those of the Passenger Car Subcommittee. This committee set the basic direction for vehicle industry policies. It issued its interim report on December 18, 1962, taking as its premise that the import of passenger cars would be completely liberalized by the end of 1964. The report covered a wide range of issues, including: establishment of mass production systems; important finance and investment issues; problems related to the increase in import of foreign cars up to the deregulation date; improving the performance and quality of domestic vehicles while at the same time lowering costs; streamlining the parts industry; controlling the entry of new manufacturers; establishing production systems, orderly distribution and sales networks, and stable financing systems; implementation of comprehensive measurements covering roads; and investigation of laws, ordinances, and tax regulations with a view to improving the automobile industry's environment.

As a follow-up to these findings, in February 1963 MITI announced new policy measures to be given form in its International Cooperation Bill. The original name was deemed to be too much to the point and was subsequently changed to the "Special Industrial Promotion Special Measures Temporary Act." This bill was presented to the Diet on March 22, 1963. Among the industries that were not internationally competitive, the bill designated three: passenger cars, special steel, and petrochemicals. It also set down special measures for promotion of corporate mergers, specialization and concentration of production, and financing and taxation, exempting these designated industries from the Antitrust Law in these areas. The bill was nothing more than an expression of policy measures for industrial guidance, coordinating the efforts of government and private enterprise. However, once it was presented, it was seen as a direct attempt to revert to the prewar situation, when the government controlled industry and stripped corporations of their independence. It was generally felt that movement in this direction would inhibit the development of promising companies and industries by hampering their entrepreneurial creativity, which was based on the principle of free competition.

The bill did not receive the backing of important circles, and the draft did not pass in July 1963. It was dropped in the next Diet session. (For details, see *The History of the Japanese Automobile Industry*, Japan Automobile Manufacturers Association, 1988.)

Until the beginning of the 1960s, MITI failed to recognize that years of intense domestic competition had made Japanese industries, including the auto industry, strong enough to thrive in an open economic system. MITI held fast to its mistaken notion that Japanese industry had to be protected by the government. The ministry was actually just shadowboxing with itself. In light of subsequent developments, it is clear the collapse of this bill was, in the end, a true blessing for the Japanese auto industry.

Manabu Sasaki, then director of the Automobile Section (currently president of Indonesia LPG), recalled that when the bill was under discussion, Soichiro Honda was asked to halt production of four-wheel passenger cars and Honda agreed at once. Thus, if the bill had actually been enacted, Honda Motor Co. would have been prevented from producing passenger cars — perhaps there would have been no Honda Civic or Accord! Sasaki also recounted how he and representatives of automakers worked together with Diet members and financial institutions to get the bill passed. This probably reflects a lack of confidence in their ability to survive on their own if they had to confront international competition. They failed to realize that an industry's vitality is rooted in freer competition, not greater controls by the government.

II-2. Grappling with Emission Control Measures

Los Angeles was already covered in a hazy purplish smog during the late 1950s when I resided there. On some days the smog irritated our eyes. Rain was sparse in southern California; the sun should have shone brightly in a blue sky. Actually, it did, outside the city, but Los Angeles itself was enveloped in a dirty veil throughout the year. I was told at the time that the smog concentrated in Los Angeles because of the southern edges of the San Bernardino and San Gabriel mountain ranges, which separate the city from the desert in the northeast. These relatively small mountains allegedly served as a barrier and caused the air in the upper atmosphere to stagnate. It was later established that smog is caused by carbon monoxide, hydrocarbons, and nitrogen oxide.

It was around this time that the state of California took the lead in implementing measures to regulate exhaust gases. But these steps failed to have the desired effect. The U.S. Congress made automobile air pollution subject to regulation under the Clean Air Act of 1967, and the new Clean Air Act (also known as the Muskie Act) was enacted in 1970. Under this act, pollutants were to be reduced by 90% of their 1970 emission levels. Hydrocarbons and carbon monoxide emissions were to be reduced by this amount by January 1975, with nitrogen oxide emissions to follow by January 1976 (this deadline was later extended two years). Thus, pollutants were to be reduced to 10% of their 1970 levels.

The skies over Tokyo remained blue until the beginning of the 1960s. It was thought that the smog over Los Angeles had nothing to do with Japan, which was cut off from the rest of the world by seas on all sides. However, by the middle of the 1960s the center of Tokyo, too, was sometimes (depending on the direction of the wind) covered in smog. The ongoing environmental debate in the United States was attracting attention in Japan, and citizen action groups were becoming more popular. Japan's Environment Agency, established in 1971, proposed that emission control regulations in line with the Muskie Act in the U.S. be implemented during the same time frame laid down by the U.S. government. A two-stage approach until 1978 was established to control nitrogen oxide emissions. At the end of the second stage, these emissions were to average only 0.25 g/km. This figure was lower than that set by the U.S.

At the time of the enactment of the Muskie Act, both American and Japanese manufacturers felt it was impossible to achieve the established figures within the designated period. Indeed, they were not even sure how to proceed. However, if a company could not achieve these goals, it would be locked out of both the Japanese and American markets. Naturally, everyone worked around the clock to develop a car that complied with the standards. Toyota assembled an engineering team, working in coordination with companies such as Toyota Central Research Laboratory and Nippon Denso Co. Based on the results of their efforts, we were able to meet the target values by creating a system composed of a tertiary catalytic converter, an oxygen sensor, and an electronically regulated gasoline injection device. In 1977, one year before the deadline, we started selling a car which complied with the standards. In this way we were able to continue exporting autos to the American market.

The trials and tribulations our engineers went through before succeeding are difficult to recount. At one point in the development stage, they finished a prototype that conformed to the 1978 standards. I was asked to give it a test-drive. Unfortunately, the car had no power. It felt as if it accelerated with the hand brake on. We were in a real trouble for future cars. A staff member in the Product Planning Office remembered my nasty comment that the cars we made before strict pollution standards could be "golden eggs." It's a bit late now, but I wish I could eat my words.

In the end, our engineers found ways to meet the standards without sacrificing running performance. It seems obvious now that these emission reduction measures could not have been achieved at such an early juncture in Japan without the influence of the United States. Congress, the bureaucracy, and large corporations do not work in concert in the U.S.; rather, Congress and the government tend to act as a restraint on large corporations, taking firm action to protect the environment. Enactment of the U.S. Clean Air Act created a precedent that made it possible for Japan to draw up its own version. The U.S. also developed ways to test the devices controlling exhaust emissions and had established Environmental Protection Agency (EPA) standards, so it was possible for us in Japan to take action quickly. Without the initiative taken by the United States, Japanese legislation would never have established standards so strict they seemed impossible to achieve in time. American democratic activism promoted Japan's environment protection.

Added to these technological breakthroughs was the cutting down of exhaust emissions by the creation of a supply system for unleaded gas. (Leaded gas blocks the catalytic function.) In the United States, starting in 1969, Edward E. Cohl, the president of GM, vigorously promoted the use of unleaded gas because he knew that such a fuel, in combination with a catalytic converter, would effectively cut down on emissions. His foresight contributed greatly to resolving the emissions problem in the U.S.

In May 1970, it was reported that people living around the Mihara interchange of Route 7 in Tokyo had high levels of lead in their blood (later it was discovered that the measurement methods were suspect). Newspapers reported that these high levels were caused by leaded gas, and these news reports provided the impetus for the establishment of a citizens' movement urging a switch to unleaded fuel. The production of unleaded gas was begun in February 1975, and the entire distribution system was converted to unleaded gas by April 1977.

This flurry of activity came at almost the same time as Cohl's call for the production of unleaded gas in the United States. It was pure coincidence that, in the area of unleaded gasoline, Japan moved forward without following the U.S. lead. However, I admire Cohl's insight. At an early stage he had seen the need for the catalytic converter and called for unleaded gas.

Since the latter half of the 1980s, other measures directed at various sources of air pollution, including the automobile, have been pushed forward. Once again you can see the blue sky over Tokyo, and smog alerts are issued only a few times a year. The area that still needs to be tackled is nitrogen oxide from diesel engines.

II-3. Look-Alike Cars: The Hidden Key to Success or Failure

In July 1967, president Kamiya announced the establishment of an additional sales channel for popular cars (the first was our longstanding Publica dealers channel). This second sales channel was to be called "Auto dealers." His announcement caught us by surprise. We suddenly had to come up with a new product plan. Luckily, the announcement coincided with minor changes for the Corolla which were to be made the following year. We pushed ahead with these changes, bringing out the Corolla-Sprinter as a two-door sporty coupe version of the Corolla. As a stopgap measure we drew up a product development plan to sell the Corolla-Sprinter only at the new sales channel, which would also handle the Publica, while selling the Corolla (except the Corolla-Sprinter) and the Publica at Publica dealers.

This move by president Kamiya was based on a decision that the mass market was too big to be handled by just one sales channel. Even though he realized that the whole plan was rather unrealistic, since it took more than five years to get a product ready for the market, he went ahead with establishing a new sales channel. This decision placed an enormous pressure on product development staff. When establishing new sales outlets, we couldn't forget the failure (mentioned previously) of our attempt to nurture small-scale stores when we had set up Publica dealers. Since we were going to establish large-scale dealers, we needed a large amount of capital. This condition meant that many potential dealer sites had backing from existing outlets. In essence, we were developing a new business, but the endeavor in many cases was dependent on veterans who already had experience in automobile sales management. Consequently, they presented us with a long wish-list of the types of products they wanted developed, to ensure the growth of their new sales channel. As a result, the Corolla, the mainstay of the Publica dealers, would also be sold at the Auto dealers. The new Auto dealers opened for business in May 1968.

However, the Publica dealers strongly protested that these measures and actions infringed on their sales rights, particularly at a time when it appeared that they would be brought out of a long difficult period through the success of the Corolla. President Kamiya stepped in and told them, "We'll change the name of the Publica dealers to Corolla dealers. You shall deal exclusively in Corollas, and shall stop selling the Publica. The auto dealers will concentrate exclusively on the Corolla-Sprinter model, which will be separated from the Corolla together with the Publica. We will leave it up to you whether you want to continue to sell the Publica and become an Auto dealer, or whether you want to deal exclusively in Corollas and become a Corolla dealer." All the former Publica dealers, with one exception, became Corolla dealers. The new sales system was then launched in April 1969.

This was when we ran head-on into the problem of "look-alike cars." Because of the way the Corolla and the Sprinter were made, they both had to be constructed on the same platform; this kept down manufacturing costs and limited the number of development processes, but we ended up with cars that, well, looked alike. Only such items as the grille, taillights, hood, and front fender were different. What we really needed to do was to change the configuration of the doors, but that would, for all intents and purposes, entail changing the entire body. This was close to impossible for a wide variety of reasons, such as the great number of preparatory processes involved in design and production. With two cars like fraternal twins, the problem was which car to name "Corolla" and which to name "Sprinter." As always, the final decision was made by president Kamiya.

The sales outlets were, of course, unhappy about the lack of differences between the two cars. As humans are apt to do, they coveted the car they did not have. The relationship between the two lines of dealers was marked by rancor and resentment.

President Kamiya stepped in and said, "Even though Kawahara's still something of a bureaucrat, he's the only one for the job." With that I was moved from the Toyopet dealers to a position in charge of the Corolla dealers. Kamiya said he wanted me to develop the strongest and largest channel in Japan for Corolla dealers, modeled after the sales channel GM had established for the Chevrolet (this was the "strongest and largest" auto sales channel in the United States). With this as my mission, I asked dealers to set up new sales outlets and increase their sales personnel in order to catch the wave of rapid and promising growth in the Corolla market. The sales outlets also believed, based on their own market information, that the future was rosy, so they were happy to cooperate. We started breaking sales records on a regular basis. By 1970, we had sold an accumulated total of 240,116 Corollas. The Corolla had overtaken the Corona as the top-selling car in Japan. Then, starting in 1974, the Corolla surpassed the Volkswagen as the number-one-selling car in the world. And so it was that we created the largest and strongest auto sales channel in Japan.

Other sales channels were jealous of what was happening with the Corolla dealers. Some heads of sales outlets went directly to president Kamiya and asked him to remove me from one of my positions. To them it was unjustifiable that I was in charge of the Product Planning Office while also acting as department manager of one of the sales channels. The suspicion arose that my sales channel was being favored when developed products were distributed. This was not true. I had no decision-making power. Decisions as to what car was to be sold by which sales channel were made by the top executives, namely Kamiya and Toyoda. Kamiya, of course, was not moved by the entreaties of the other dealers.

Let's return to the problem of the look-alike cars. Starting in 1980, to complement the Mark II, Toyota introduced two look-alike models into different domestic sales channels: the Cresta and the Chaser. These models had the same platform in common. There were hardly any obvious differences, except for pressed doors on one of the models. The sales outlets that had just started to handle the Cresta and the Chaser had to find new customers who might not be attracted to the more commonly known Mark II. Toyota was lucky to expand its share in this dramatically growing market. We were also able to put pressure on Nissan in this same market.

Nissan had introduced five different models in this market: the Laurel, Skyline, Leopard, Maxima, and Cefiro. The platform for each car was different. This helped create a unique lineup, but Nissan's market share did not rise at all. The reason for this failure was clear. If you try, as Nissan did, to put out so many cars in the same class, it will be impossible later on to make model changes for each of them during the same period. Also, it is inevitable that some cars will have more appeal than others. People want the latest, the best. Although you offer them a lot of autos, demand will concentrate on only a few. Nissan customers spent all their time trying out different cars, and Nissan didn't have time to go looking for new customers.

Even today Nissan offers many more models than Toyota, but its total sales volume is far lower. Trying to sell look-alike cars comes with its own set of problems for both the manufacturer and the dealer, but Toyota's strategy has proven to be better.

In 1977, when GM decided to downsize its passenger cars, the Cadillac, Oldsmobile, and Buick all began to resemble each other. Ford embarked on a PR campaign trumpeting this fact. Cadillac, Oldsmobile, and Buick had each nurtured its own special customer base over a long period of time. Sales dropped dramatically after the release of cars that lacked personality. Customers felt jilted. This period has been pinpointed as the beginning of GM's fall from glory. Look-alike cars proved to be a failure for GM because of its long-held tradition of distinctiveness. But look-alike cars were a success for Toyota, because they were part of our process of creating a customer base.

II-4. The God of Sales: President Kamiya

At some point in time, Kamiya, president of Toyota Motor Sales, became known as the "God of Sales." This sobriquet probably grew out of respect for the performance of the hardy Toyota sales network Kamiya had built. Those of us who worked under him frequently witnessed his marketing genius. His sharpness and intuition received the admiration and respect of the employees at Toyota Motor Sales as well as management executives and engineers at Toyota Motor. For us, he truly was a god.

The Corona had won over the Bluebird. Now, together with chief engineer Atsushi Tajima, we set our sights on the next Corona model. We developed the Corona Mark II, a compact, widening the interior to match the Opel Record and giving it a 2000-cc engine. The total length, width, and height of the Corona Mark II was only slightly increased overall. The prototype was approved at a joint inspection conducted by Toyota Motors and Toyota Motor Sales. President Kamiya, who was in attendance, proclaimed that the car could thrive alongside the current Corona, and that it was our job to ensure that their production and sales were both successful. Right then and there it was decided to study the idea of continuous production of the current Corona. The importance of president Kamiya's proposal to the subsequent development of Toyota is evidenced by the trends in the foreign and domestic sales of the Mark II and Corona. ("Corona" was later dropped from the name, and the car became known as the Toyota Mark II. The export model is called the Cressida.)

During one examination of the new Publica model, people complained that it lacked style. Kamiya proposed that a belt-shaped accessory plate be added. The designers immediately gave the proposal serious consideration and came up with an accessory plate that gave the car a look of strength and elegance. This plate was subsequently dubbed the "Kamiya belt" by the designers and was put on all models thereafter.

Examples of Kamiya's wisdom abound, including his deft handling of the problem that arose between the Publica dealers and Corolla dealers and his decision to establish the Auto dealers. These events illustrate his uncanny marketing intuition.

Kamiya also had real charisma. Kiichiro Toyoda, later the president of Toyota, who started truck production, met Kamiya when he was serving as manager of the Advertising Department of GM Japan. He felt real respect for Kamiya's character and admired his abundant sales experience. At their first meeting he asked Kamiya to join Toyota and told him he would give him full responsibility for the company's marketing. Kamiya, too, was impressed with Toyoda's character and his passion for establishing domestic car manufacturing. He accepted the offer on the spot. He resigned from GM Japan and was appointed general manager of Toyota's Sales Department. This was back in 1935.

At the time, Ford and GM controlled the Japanese auto market, and few domestic autos could compare with the imported assembled vehicles in terms of quality and cost. So it was not easy for Kamiya to set up a sales network. He appealed to the patriotism of targeted sales outlets, asking them to help establish a domestic automobile industry.

In 1935, a Buick dealer changed over to Toyota, creating its first dealership in Nagoya (the dealership is currently Aichi Toyota Co.). The following year, Toyota dealerships were established in the prefectures of Tokyo, Osaka, Shizuoka, Hiroshima, Gifu, Mie, and Tochigi. By the end of 1938 a total of 29 dealerships had been opened nationwide. Most of the dealers had moved from GM to Toyota. Toyota thus established a franchise system that nearly covered the entire domestic market.

All these sales outlets had been drawn in by Kamiya's character and passion. It was he who had induced them to quit dealing in foreign cars and join the Toyota sales network. But these dealers, once established, were in for considerable difficulties. Kamiya had learned the American way of monitoring and supervising sales outlets during his days with GM. Therefore, when Toyota adopted GM's franchises, it also took on much of GM's dealer management system. Kamiya had sales outlets report every ten days on orders received and number of units sold, like the GM Ten Day Report System. He also aimed to standardize items in the balance sheets and income statements of dealers, like those at GM, and drafted a unified accounting manual. Management was strict in areas such as honoring payment deadlines for bills.

However, Kamiya was very critical of GM's approach to the relationship between manufacturer and dealers. When sales outlets got into trouble, GM would cut them off without compassion, and their sales rights would be transferred to another dealer. From the standpoint of the Japanese business style, which strives for peaceful coexistence and mutual prosperity, this approach was cruel and heartless, hardly the way to build a strong sales network. Kamiya therefore made significant modifications in management's approach so that if a sales outlet found itself in trouble, it would be helped back on its feet with special financing or personnel from Toyota. Sales personnel formed a strong bond of friendship with Kamiya that went beyond business, even playing bridge with him — a game he liked very much.

What eventually developed was what could be called the "Kamiya Doctrine." This can be summed up as "Customers come first, the dealers come second, and the manufacturer comes third." In other words, profits from the sales of autos should first be passed on to the customer, then to the dealers, and lastly to the manufacturer. This approach resulted in the company's gaining customer trust and sales outlets flourishing. These benefits, in turn, helped the manufacturer develop.

Based on this doctrine, Kamiya's pricing policy stated that, in order to get a lot of drivers behind the wheel of Toyota vehicles, we had to offer cars that at the very least were cheaper than foreign ones. He recognized that vehicles made by Toyota did not yet measure up to foreign cars in terms of quality. His policy was to make up for that through post-sale service. It was then that the Toyota tradition of focusing on service was born.

With the advent of the World War II, truck sales were put under a distribution ration system, with sales outlets being gathered under a distribution control company. At the end of the war, this company was disbanded, and manufacturers turned their attention to rebuilding sales companies. president Kamiya set out to restore prewar Toyota dealers. He suggested to obviously talented staff of other dealers, particularly Nissan, that they join Toyota dealerships. Here, too, Kamiya's doctrine had a positive effect. By around 1948, all the prefectures in Japan had Toyota dealerships. Moreover, Toyota's sales channels were dramatically stronger, both in quantity and quality, than in prewar days.

In the spring of 1953, Kamiya embarked on a surprising experiment at Tokyo Toyopet, a Tokyo-based dealership, that proved a great success. The experiment was to employ new college graduates as sales personnel. College students who were interested in the auto industry and wanted to enter Toyota Motor Sales and Toyota Motor were rapidly increasing in number. However, only a limited number of them were able to join automakers; the remaining were forced to work in other industries. President Kamiya's idea was for Tokyo Toyopet to employ some of them. A majority of the college graduates who were employed by dealers were those were unable to get a post with an automaker. Some of them wanted to join Tokyo Toyopet right from the start, because they thought they would be able to stay in Tokyo if they were employed by the company. If they joined the government, financial institutions, manufacturing companies, or trading houses, they would have to be prepared to leave Tokyo, possibly being ordered to transfer to branch offices in other cities, even overseas. Not a few of these new employees were from wealthy families, which helped the company put down roots in that market. One of Tokyo Toyopet's major strengths today is that all its top executives have had sales experience.

Sales of the ToyoAce, introduced in 1954, did not grow as initially expected, although production facilities for the vehicle had been expanded. In January 1956, president Kamiya asked officials of the Toyota Automobile Dealers Council to agree to establish a new sales channel to handle mainly the ToyoAce and the Corona. He stated, "A one-liter container can hold only one liter of water. If you want to store two or three liters of water, you have to increase the number of containers." A network of Toyopet dealer channels was established across the country by July 1957. Toyota subsequently increased its number of sales chains as the market expanded. Now the automaker has five sales channels: of Toyota, Toyopet, Corolla, Auto, and Vista.

Following Toyopet's example, some of these Toyota dealers started hiring young college graduates as sales personnel. Today, a total of 2,000 to 3,000 college graduates are employed by Toyota dealers every year (they account for 50% to 60% of all college graduates who join the auto industry). The employment of such a large number of graduates is one of the secret sales strengths of the Toyota group.

Many Toyota dealers, who developed in the manner described above, have been named top-rated companies in each area. They also contributed to their local communities, promoting the development of local industries and acting as business leaders, in such way as serving as executives of local chambers of commerce. Some dealers who have served Toyota for a long time owe their considerable fortunes to Toyota. The strength of the sales network established by president Kamiya, "God of Sales," also derives from Toyota dealers.

Konosuke Matsushita, founder of Matsushita Electric Industrial Co., was also called the "God of Sales" by the mass media at that time. Japan has been blessed with two "Gods of Sales."

I was one of the many people drawn by Kamiya's charisma. He always said that it was important to think from the viewpoint of other people before taking action. Whenever he talked with Toyota dealers and subordinates, he tried to look at things from their point of view. He never used his position as president to coerce or persuade. This attitude was the major reason so many people felt drawn to him.

President Kamiya passed away on December 25, 1980, at the age of 83 (he was honorary chairman at the time). Many people felt that his death marked the end of an era.

II-5. *Overtaking the Datsun Bluebird*

I continued working in product planning at Toyota Motor Sales until my retirement in 1981, when Toyota Motor merged with Toyota Motor Sales. During those years I traveled to the U.S. and Europe to visit distributors and dealers. I attended motor shows in each area, visited the design workshop Carrozzeria in northern Italy, and was given tours of many automobile plants. Wherever I went, I drove as many cars as possible. This is how I was able to understand the differences in running performance between Japanese, American, and European cars. The speed limit in the U.S. was generally 60 or 65 mph (approximately 100 km/h). However, Congress enacted a nationwide speed limit of 55 mph as an energy-saving measure in response to the first oil crisis in 1973 (although I understand this nationwide speed limit has been lifted by some states, back to the previous 60 or 65 mph).

The American speed limit is very low in contrast to that on the German autobahn, which does not even have a posted speed limit, and that on roads in France, Belgium, Holland, and the United Kingdom, which post limits of 120 km/hr. These differences in speed limits gave rise to differences in automobile performance between U.S. and European cars.

Expressways became more common in Japan after the 1970s. Vehicles produced in Japan were designed with speed limits similar to those in the United States in mind, so they had engines that could take on American roads. However, these same cars had problems with faster European drivers. In Germany there is a correlation between a car's price and its maximum speed. Japanese cars are expensive, so their high-speed performance must be good. This, at least, is the litany I have often heard. At motor shows in Frankfurt, Geneva, and Paris, all cars are displayed with their price and maximum speed. A point of pride is the higher speeds of the higher-priced cars. Anyone driving in the inner passing lane on the autobahn is likely to be overtaken by higher-priced cars like the Benz or the Citroen DS, which come up from behind at speeds of 200 km/hr or more, flashing their headlights and moving everyone out of their way. One could say that the maximum cruising speed is the actual speed limit.

Note: European cars, both top-of-the-line makes and those aimed at the average consumer, run stable even when the accelerator is pushed all the way to the floor. While most Japanese cars have powerful engines, they cannot cruise at their maximum power because at high speeds the driver gets the feeling that the front wheels are lifting off the road surface — hardly a stable drive. Large American cars are the same. The driver must release the accelerator a little and find the optimal speed, probably just under the maximum speed. Winds on the European continent are often strong; cross-winds, even head-winds, can sometimes push a car at high speed one meter laterally. European expressways traverse rolling terrain and steep landscapes where many cuts are needed. Cars must cross wind-exposed bridges straddling ravines. Drivers have to adjust their speed to different driving conditions.

Because of the faster speeds, car features must be different. In Germany, a Celica was driven at more than 140 km in the rain. The windshield wipers hydroplaned above the surface of the curved glass, rendering them totally useless. To say that driving at 140 km/hr in the rain is lunacy is not sufficient. Some drivers barrel ahead at these speeds in snowy conditions. When halogen lamps were first put out, we were subject to unceasing requests to install them on Toyotas. It was found that just a little extra light made driving at high speeds after dark easier, but unfortunately, the company was unable to procure the lamps in time, so we couldn't satisfy these people.

Cobblestone roads, still found in many European cities, were a source of misery for Toyotas. Even at low speeds, the steering wheel would vibrate violently. We nicknamed this type of road the "Belgian road" and built a replica in our test course in Toyota City to find a way to deal with this problem.

At any rate, we were aware that even European vehicles could not boast perfect performance. For example, the 1973 BMW did not always exhibit good stability at high speeds. If it was driven over 160 km/hr in a cross-wind, it could be difficult to keep straight. The Audi 40 was not built to withstand direct winds. While this may be a bit of an exaggeration, I found that at speeds of 120 km or above it flitted across the road like a fallen leaf. The Benz 280 would take a long time to reach 200 km/hr. You had to push the accelerator down to the floor. Moreover, an unpleasant vibration would develop if the tires were even slightly unbalanced.

Meanwhile, I realized these manufacturers had been making efforts to improve their products. I drove an Audi 80 again in Germany in 1978 and discovered that its handling performance had gone up compared to a couple of years before. Immediately afterward, I went to the Geneva Motor Show. One of the managers at the Audi booth told me, "Since Ferdinand Porsche II became a vice president of the company, we have made every improvement he suggested. Audi has improved to such an extent that it is nearly unrecognizable." Around the same time, I drove an Audi 100. It was powerful, but truly frightening at speeds above 170 km. Dr. Porsche's reach may have not extended to this model. It was put on the American market under the name Audi 1000, but it got a reputation as an unsafe car, and sales did not meet expectations.

Toyota improved as it survived the rigorous technical competition that existed in the European market. By the mid-1970s, Toyota was appreciated as a durable car not likely to break down. Our "people's car," Corolla, measured up to European vehicles.

However, in the Cressida class (the name of the Mark II export model), we continually heard complaints that speed performance did not meet the price ratio. This auto was dramatically improved in the beginning of the 1980s and could be driven with the accelerator all the way to the floor. It has been said that the advice of engineers working for the British Lotus, which had signed a technical cooperation agreement with Toyota, was extremely helpful in making these improvements. But going back to the Audi story, it is extremely interesting to note that even at the end of the twentieth century, there are still areas of auto engineering that are beyond the reach of scientific analysis, and a lot still depends on the intuition and experience of well-trained engineers.

Note: Toyota obtained a share of about 23% of Lotus after 1975, establishing a technical cooperation with the company. Toyota supplied Lotus with engine parts and received technical guidance in many areas. In 1986, GM embarked on a buyout of Lotus, at which time Toyota gave up its shares in the company and terminated the technical cooperation agreement.

At the end of the 1970s, Toyota embarked on the development of a luxury-class, high-performance coupe. Our goal was to create a car that could compete on a par with Benz at speeds of 200 km or higher on the autobahn. A monthly sales target of 3,000 units was set, and the car was named the Soarer. This endeavor was revolutionary, not just in its pursuit of high-speed performance but also because it signaled Toyota's decision to go for small-lot production of a car for a limited market. When the fully developed prototype came out at the beginning of 1981, I went to Europe to participate in the test runs. With great anticipation and pride I started to bring the car up to speed. It ran well up to about 180 km/hr, but once over that speed, the front end lifted off the pavement. The car was, I had to admit, unsteady. When it passed a large truck at high speed, it was pulled toward the truck; when it was beyond the truck, it was pushed over to the opposite side. The Soarer clearly did not run on a par with the Benz. Lack of stability at high speeds did not present a problem in the domestic market, because speed limits kept driving sedate. So sales were steady. However, we had to wait for the launch of the Lexus in 1989 before a Toyota could match the performance of a Benz at high speed.

II-6. The Lexus

Toyota introduced its luxury car, the Lexus LS 400, on September 1, 1989. Before this (May 1989), TMS (Toyota Motor Sales, U.S.A., Inc.) had invited nine well-known American journalists to Germany to test-drive the Lexus prototype as well as the Benz, BMW, Audi, and Jaguar. The participants were given free rein on the autobahn, which has no posted speed limit. All the journalists were asked not to publish anything about their test drives until the market release of the Lexus. When August rolled around, all the writers reported that, based on their test-drives, the Lexus was quicker, quieter, and offered a more comfortable ride in comparison to the best the world had to offer.

Right after the Lexus hit the market, sales rose steadily. The typical buyer was someone who had previously purchased European luxury cars (cf. *Against All Odds* by Yukiyasu Togo and William Wartman, St. Martin's Press, 1993). The Big Three, as well as European luxury car manufacturers, found it hard to believe that the Lexus was so good until they actually got behind the wheel.

In an article entitled "The Big 3 Have Lost Their Customer Base" in the *Financial Times* (June 25, 1989), Richard Fest stated that engineers from one of the Big Three got his hands on Toyota's new Lexus at the end of 1989. Until this point they could only go on what they had read and heard about the auto. Once they saw the real thing and fully comprehended the creativity, effort, and capital that had been poured into the Lexus, they were extremely discouraged. They also realized that what they had done until then did not match the Lexus. Big Three executives began to study what they needed to do to come up with a vehicle to take on the Lexus. This was tantamount to admitting they had lost their market base. It had been accepted without question until then that American manufacturers were best at building large-sized luxury models, while the Japanese produced the best compacts. Now they saw that Japan could do both.

Note: Lexus was the name of one of two sales distribution channels under the TMS umbrella (the other was called Toyota). The Japanese version of the Lexus was called the Celsior, while the American version was dubbed the Lexus LS 400, which was the name of its sales distribution channel. It is referred herein simply as the Lexus.

According to *The Road to the Lexus/Celsior* (Osamu Nakajima, Diamond Co., 1990), after the Lexus hit the American market in 1989, GM bought 14 of them, Ford bought eight, while Chrysler got two. The Lexus went on to outdo all American and Germany luxury cars in the Initial Quality Survey (IQS) released by J. D. Power in May 1991, the Consumer Satisfaction Index (CSI) released at the beginning of July, and the Sales Satisfaction Index (SSI) at the end of July. These plaudits all attest to the favorable reception the Lexus received among drivers. The April 1992 issue of *Consumer Reports* named the Lexus the benchmark for luxury cars.

Ichiro Suzuki (then the chief engineer in charge of the development of the Lexus, currently Toyota Auto Body's managing director) and Kazuo Okamoto issued a report entitled "The Toyota Celsior" in the *Journal of the Society of Automotive Engineers of Japan* (No.3, 1990). The article briefly explains the Lexus engineering development process. The two authors start out by stating, "We condensed Toyota's 50 years of engineering experience into developing the Toyota flagship car as a way to push the company into the 21st century." They describe their development goals as follows: "In developing the Lexus/Celsior, our aim was to create a high-performance car of world-class luxury which purred with more than a touch of human warmth at low, medium, and high speeds." A high-level function and real human warmth were essential to creating a vehicle. Just owning one and getting behind the wheel gives the car owner real pleasure, pride, and, as time passes, a sense that this is the car one wants to be with for a long time. Actually, high performance and warmth can be reciprocal in nature. When they are, they should coexist at a high level. This was our engineers' approach in the development of the vehicle.

A basic vehicle development concept alone is inadequate. One also needs a basic engineering development policy to achieve the aforementioned aims. "Go to the source" measures and our "yet" concepts were developed as a result. "Go to the source" measures were based on the notion of returning to the root of a problem to eradicate it. For example, in order to overcome vibration and noise, we went back to their root elements and reduced the vibration force that was its ultimate cause. The "yet" concepts were based on the notion of creating a car without compromising. For example:

- Elegant styling yet superior aerodynamics
- High performance yet low fuel consumption
- Super-silent yet low weight

While it may be a little technical, I would like to quote two sections of this report to illustrate the effort put into development.

Aerodynamic features

Considerable effort was directed at minimizing air resistance, to make it possible to cruise at super-high speeds yet with low fuel consumption. We achieved a drag coefficient CD value of 0.29, the top for a vehicle in this class. Thus 95 horsepower is sufficient for a car running at a speed of 180 km with a CD value of 0.29, while 110 horsepower is required with a CD of 0.35. In the same manner, if fuel consumption at a speed of 110 km/hr is used as an example, we obtain a difference of 4.5%.

In bringing the CD value below 0.30, it was important not only to improve the car's upper body but also to reduce obstacles to airflow under the floor and create a smooth flow. Engineers therefore made the floorboard flat and considerably reduced air drag under the floorboard by a large engine undercover, spats in the front of the tires, and fairing for the rear lower arm. Furthermore, it was essential to flatten the floorboards in order to reduce the CL (lift coefficient), which is described below.

The CL, and in particular the CLF (the CL of the front wheel area), have a major effect on linear stability at super-high speeds, and the CLF of the Lexus/Celsior, with a spoiler mounted, is minus 0.06, which is a value close to that of a sports car like the Porsche.

NVH (Noise, Vibration, and Harshness)

While reducing the car weight, we introduced a mechanism for reducing noise and vibration at their sources when the car was driven at extremely high speeds. For example, we developed a straight driving system without joint angles. Unbalanced vibrations in each rotating part were also reduced to less than one-third of previous levels for the propeller shaft. This technique considerably reduced indistinct noises.

The above excerpts from the report written by Suzuki and Okamoto underline what Toyota aimed to achieve: a car that could be driven at a high speed on the speed-limit-free autobahn.

As mentioned in the previous section, Toyota started developing the Soarer in the latter half of the 1970s, aiming for a high-performance car equal to Benz and BMW cars. However, as of 1981 the Soarer was not able to match a Benz or BMW in terms of high-speed driving stability. Monthly sales of Benz and BMW autos were increasing in Japan at the time and German cars were gaining a foothold as prestige cars as well. Popular opinion held that people who drove up to first-class hotels and department stores in a Benz or a BMW were viewed as good customers and greeted with every courtesy, while those driving up in Japanese cars had to fend for themselves. On the highway, Japanese cars were often forced to make way for German cars coming up from behind. This state of affairs was intolerable for Toyota employees, particularly the engineers, who had helped their company rise to become the third-ranked automaker in the world behind GM and Ford.

All Toyota engineers thus supported the goal of developing a car which drove better than a Benz or BMW at high speeds. When Toyota launched the Lexus project in 1983 without any fixed schedule for the start of production, all engineers devoted themselves to achieving this goal. Some voluntarily stayed overnight for many days at the high-speed wind tunnel, creating a body shape that reduced the CD value to less than 0.30. Others tried to develop a method to manufacture and install super-advanced precision parts that could not be measured using existing measuring instruments. During this period, Toyota built its own test course in Hokkaido, where we could conduct high-speed test-driving at more than 200 km/hr. Until then, super-high speed driving tests could not be conducted in Japan.

The success of the Lexus can be attributed to Toyota's strong drive to challenge Benz and BMW, which had lorded it over us in the U.S. market. It can also be attributed to Toyota's willingness to show Americans, who had believed in the superiority of German cars, that Toyota not only made high-quality, mass-produced compact cars but also possessed technology equivalent or superior to that of European automakers.

Lexus, called Celsior in Japan, was put on the Japanese market in October 1989, one month after it was introduced onto the U.S. market. Since the success of the car in the U.S. market was much talked about in Japan, dealers were flooded with orders immediately after the official announcement. Some customers who placed an order in the first month of release were forced to wait several months.

I can recall the moment I tried out the car for the first time in the late fall of 1989. It rained heavily on the highway, but Celsior took to the road like a battleship triumphantly plowing through the waves. It ran smoothly and quietly and made me wish I had been listening to an aria on the high-quality stereo installed in the car as standard. I experienced supreme bliss for the first time in 35 years at Toyota and 45 years of devotion to automobiles.

In October 1994, Toyota remodeled the Lexus/Celsior for the first time in five years, introducing a new model into the Japanese and U.S. markets at the same time. Like European luxury cars, its look after the model change was not all that different. The total length and width were the same as those of the previous model, although the interior width was extended 20 millimeters and the leg room of the rear seat was expanded by 70 millimeters. The total weight was reduced by 130 kilograms. The power and torque of the engine were also enhanced. Noise inside the car while driving was only 62 decibels, which is incredibly low. Only those who have actually driven the car would understand what it means. It may be a bit of an exaggeration to say so, but the quietness of the car reminded me of the solemn silence inside a stone church. Its maneuverability and stability at high speeds were improved so much that I doubted there was room for improvement. Toyota was said to have received orders for 15,000 cars in the new model's first month. I look forward to further growth in the future.

Chapter III

Management of the Automobile Industries in Japan and the U.S. in the Postwar Period

Management of the Automobile Industries in Japan and the U.S. in the Postwar Period

III-1. Characteristics of the Automobile Industry

The characteristics of the automobile industry should be discussed to clarify my viewpoint on the development of the industry in Japan and the United States after World War II. The automobile industry begins with investment and ends with investment. Chart 3 shows the operation cycle. It begins with capital investment for production facilities. Then, by moving through production, sales, and collection, it can achieve operation results, or net profit. This profit is then distributed into internal saving or to outside investors. With profit saved in the company and additional outside investment, the company reinvests and continues the operation cycle.

Capital Investment

Today's auto industry can be categorized as a capital-intensive industry. Therefore it must spend enormous amounts of money on equipment and facilities. At the same time, it must also develop competitive new products every year. The scale of this R&D investment has recently become so enormous that it exceeds investment in production facilities. Other important capital investment activities, though smaller in scale than those for facilities and R&D, include investment in subsidiaries and affiliates. This includes capital participation in suppliers and component manufacturers, as well as financial assistance to dealers to solidify sales networks. Other investments involve procurement of raw materials, payment of wages, and raising of working capital such as sales financing and inventory financing.

Chart 3: Operation Cycle of Automakers

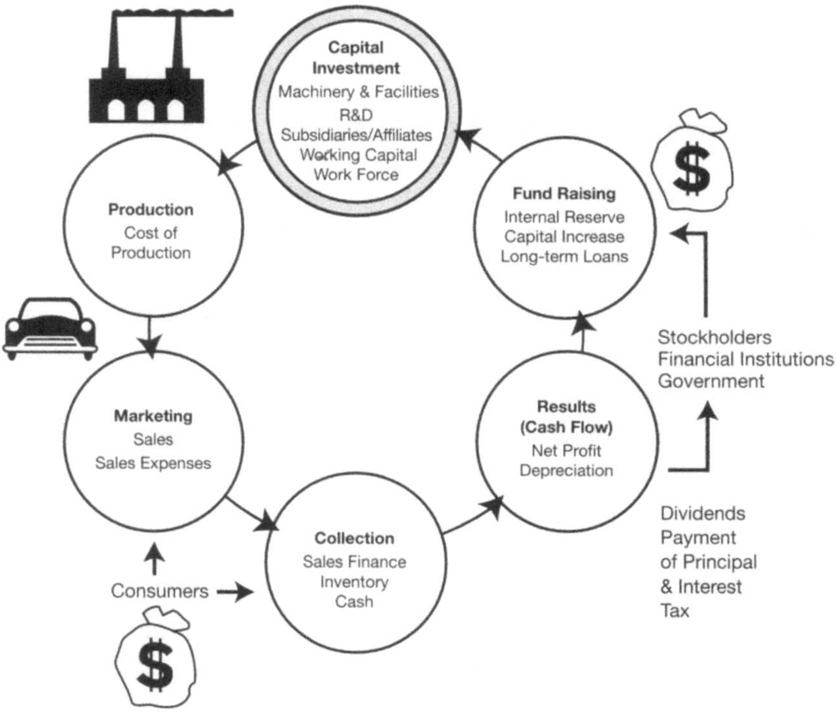

Recently, even the labor force has come to be considered a target of capital investment. The cost of educating automotive engineers and training shop-floor workers to competently utilize high-tech equipment can also be viewed as capital investment. Labor strength can also be categorized as a true asset. With these types of capital investment, plant facilities can be built, and then production is begun.

Production, Sales and Collection

With production comes production cost. The products are sold wholesale to dealers. At this stage, sales incur cost of sales. Then comes the retail level, which is the domain of dealers. Their sales activities takes various forms, such as new car sales, used car sales, installment sales, leasing sales, parts and service sales, and insurance sales. Next in line are the consumers. They make the payment, sometimes directly to the dealers or, in some cases, through the trade-in of their used cars, or through outside financing companies or those operated by manufacturers. Thus the manufacturer collects the payment from the dealers. With the sale of the car to the consumer, the production/sales operation cycle comes to an end.

Results

As the conclusion to the production/sales operation cycle, cash flow that covers depreciation and net profits is generated. From net profits, dividends and interest as well as taxes are paid. What remains is put into internal reserves.

Fund-Raising

The enterprise must again turn its cycle toward the production operation and secure the needed funds for investment. If the increase in internal reserve is insufficient for the next cycle of investments, capital is increased or a long-term loan is made. Or, if there is extra cash flow, payments are made on the principal. With this, the cycle turns over to the next production sales operation.

This revolving operation cycle starts with capital investment. The main difference between the automobile industry and other industries is that the life cycle of new products is short. Technological advancements are made at such a fast pace, and consumer tastes change as well, leading to changes in lifestyle. The automotive industry, therefore, cannot continue its business unless it keeps developing new products or improving existing ones, always striving to compete with other automakers in supplying products that appeal to consumers. It is the responsibility of the top management of an auto company to decide on the development of these products and to secure the investment necessary to purchase equipment. This cycle, which starts with capital investment and ends with investment, is characteristic of the auto industry. And competition is the force that drives this operation cycle.

III-2. *The Evils of the Big Three's Oligopoly*

The market share of the Big Three had already reached close to 90% by the latter half of 1930s, before World War II. They further boosted their share to 97% by the late 1960s (American Motors Corporation held the remaining 3%). GM steadily raised its share from around 40% in the early 1930s and acquired the lion's share with 51% by 1954 and 52% by 1962. The U.S. automobile industry has thus been dominated by the Big Three, led by GM, since the late 1950s. Under this oligopoly, the three auto manufacturers did not compete with each other. Instead, they gave priority to their own profits and ignored customers' needs. The following list includes some examples of the harmful effects of the oligopoly.

(1) GM decided to calculate the price of its cars by adding target profits to its cost. The other two makers followed suit, choosing to secure stable profits rather than to challenge GM's pricing policy.

(2) The Big Three tried as hard as possible to avoid making improvements in their autos that would lead to a cost increase. Japanese auto manufacturers, which have faced severe competition, often give the highest priority to the improvements of their autos regardless of the cost, and think of cutting costs later.

(3) The management goal of the Big Three was to gain maximum profit from minimum investment. As a result, their production facilities aged, and their cars became obsolete.

(4) Improvements in auto performance, safety, and fuel efficiency were delayed. The Big Three were reluctant to introduce environment-friendly technologies.

(5) They did not try to collect information about users of their cars after sales and made light of or ignored the results of consumer surveys.

One glance at the special annual issue on cars published by *Consumer Reports* is enough to confirm that the quality of American cars is far inferior to that of imported cars, in particular Japanese and German cars. Such data have been available to the public for 20 years. Surely American car owners have been irritated by the inferior quality of their cars. U.S. auto manufacturers, which have made enormous profits by selling such low-value automobiles, and the United Automobile Workers (UAW), have negotiated extraordinarily high wages to produce them. Their gain came at high cost to their customers.

On April 11, 1991, the chairmen of the Big Three — Robert C. Stempel of General Motors, H. A. Poling of Ford, and Lee Iacocca of Chrysler — appeared on a special news program on ABC's *Nightline* entitled "Made in the U.S.A.: American Cars." During the program, Iacocca complained that captive imports (Japanese-made cars sold under U.S. auto manufacturers' brands) did not sell as well as Japanese brand cars and that U.S. consumers did not buy American cars although their quality and performance had improved. Iacocca apparently hadn't gotten the message that American consumers had lost confidence in the Big Three automakers. American consumers simply wanted good value for their dollar.

It is natural to think that the lack of competition — or nearly no competition — among the Big Three allowed General Motors to persist in its target return pricing (the setting of prices by adding costs to the target profits) until around 1970. In the U.S. car market, competition existed only among franchises of the same dealers. It was often said at that time that the only rivals of Ford dealers were Ford dealers and that the only rivals of Chevrolet dealers were Chevrolet dealers. The Big Three kept increasing the number of dealers and pushed ahead with hard-pressure sales tactics to boost sales at the expense of their own dealers.

Note: Kiyoshi Yamazaki explained General Motors' pricing mechanism in his book, *Automobile Power in Japan, the United States and Europe* (Diamond Co., 1971). In it he wrote:

> *The cost of a car is calculated by estimating production costs for the standard production volume based on the average operating rate of 80% and adding standard sales expenses (which account for 7% of sales) to the production costs. The standard price on the product is determined by adding the target profit calculated based on a target profit ratio, to the cost price. The target profit is called an optimum profit rate in General Motors and it is equal to 20% of the return on equity or 15% of the total liabilities and net worth.*

As mentioned in the previous section, the business operations of auto manufacturers developed through the operation cycle, which begins with investment and ends with investment. The oligopoly of the Big Three, however, prevented the cycle from coming full circle. Most of the profits they made were flowing out of the industry, and reinvestment in products and facilities was ignored. They fell into financial difficulties after the energy crises, because during the period in which they failed to reinvest their profits, their products had deteriorated, their production facilities were outdated, and their quality and productivity had become inferior to those of Japanese auto manufacturers, who got a later start. Therefore, imported cars steadily penetrated the market. The Big Three have repeatedly accused Japan of its unfair competition, but these hardships were incurred by their own indulgence in oligopolistic practices.

III-3. Comparison of the Management Performance of Toyota and General Motors in the Postwar Period

The Japanese automobile industry grew rapidly and put pressure on its U.S. counterpart in the 20 years from 1960 to the end of the 1970s. Japanese manufacturers, which had been almost ignored by U.S. automakers, surpassed their U.S. counterparts in quality and productivity in an extremely short period of time. Japanese automakers then penetrated deeply into the U.S. car market, dominated by the Big Three, and grabbed more than 20% of the market. This growing presence of Japanese makers in the U.S. market is now viewed as a major cause of the trade imbalance between Japan and the United States. In this section, the root of Japanese-U.S. automobile issues will be briefly analyzed by comparing the business performance of Toyota and General Motors during the period in question, in light of the characteristics of the automobile industry mentioned in the previous section.

Note: I gave lectures on the automobile industry at Hitotsubashi University from 1981 to 1982, using the materials that Kenji Suzuki (a staff member of Toyota Motor Sales' Research Department at the time, now manager of Toyota's Product Planning Department) collected from the automakers' annual securities report and General Motors' annual report, based on the advice of Terumasa Hirano, a U.S. CPA who is now president of Springer-Verlag Tokyo, Inc. after working for the Arthur Young accounting firm.

Chart 4: Sources and Application of Funds

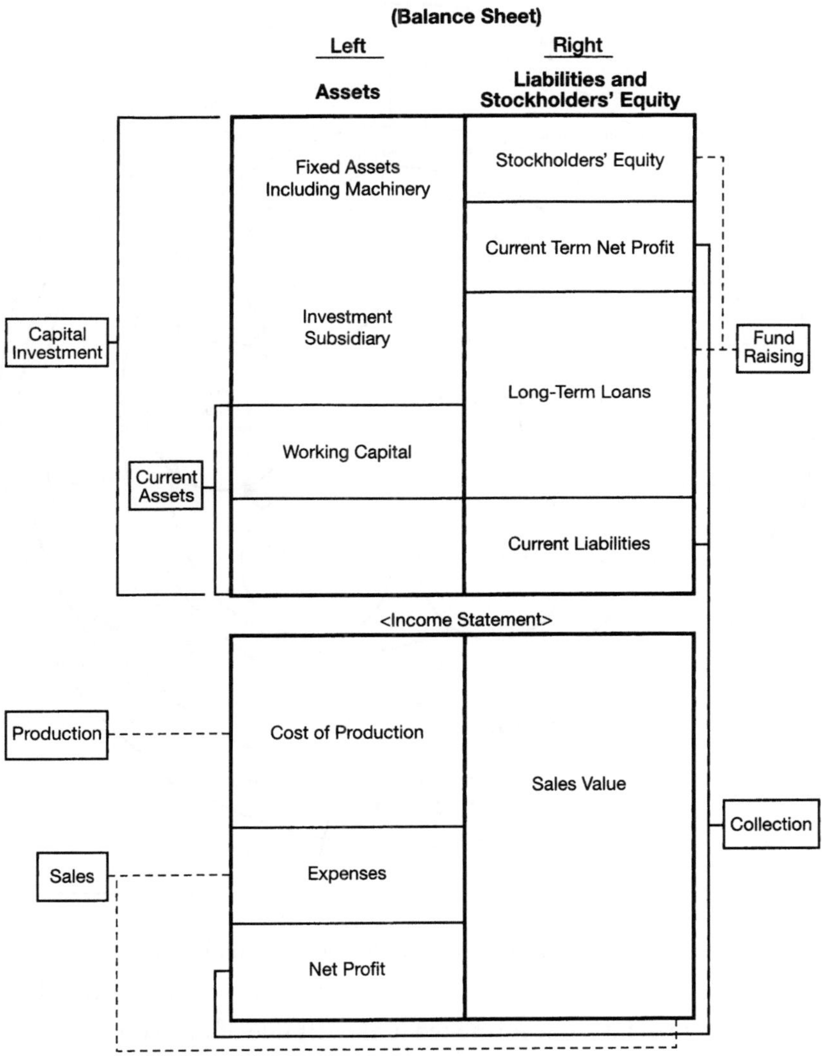

Chart 5: Cycle of Fund Applications

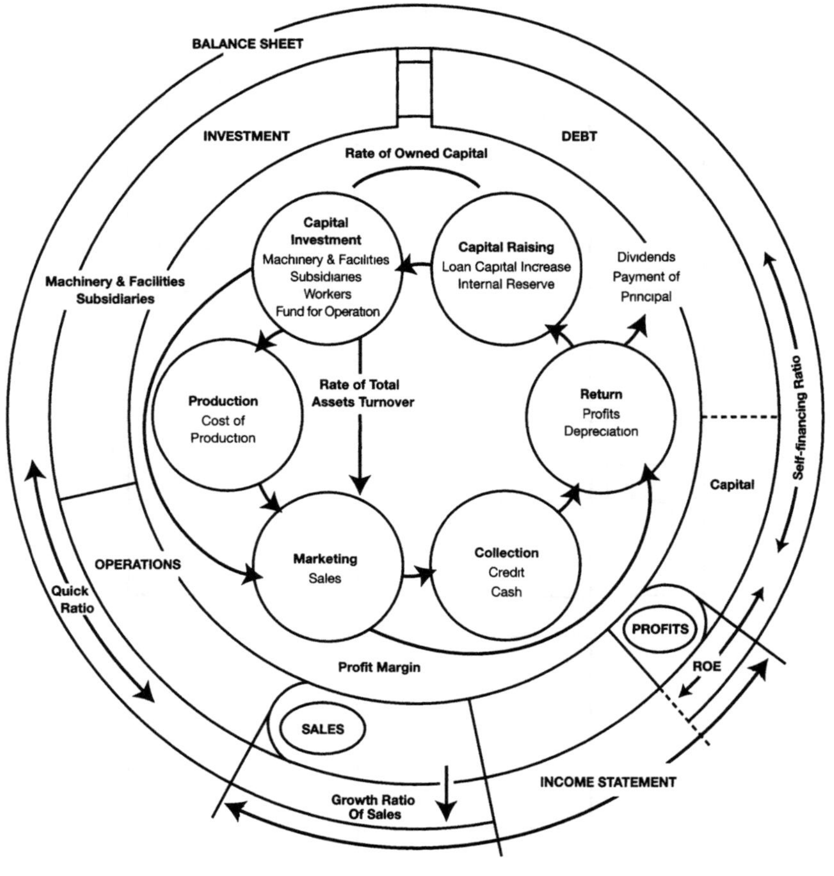

Chart 4 shows the sources and applications of funds. The upper part of the figure is a balance sheet, the lower one an income statement. The content of capital investment is stated in the left side of the balance sheet.

Production costs are incurred when production starts. Marketing activities require expenses but yield sales at the same time. Then net profit is collected and incorporated into current term net profit. Funds from the current term net profit are reinvested in production.

Chart 5 shows the cycle of fund applications. In the figure, retail sales are integrated into wholesale sales, but the same figure can be drawn for the retail trade. Many managerial indexes, which are useful in comparing business activities of companies, can be obtained from the figure. Those indexes include productivity, which shows how many sales and profits are recovered from invested capital, and turnover of total capital, which indicates how well invested capital is utilized.

Comparison between Toyota and General Motors is made below, based on various indexes. The growth rate of sales, which expands the fund application cycle, will be examined first.

Growth Potential (Growth Rate of Sales in Value)

As the growth of sales in value is such that it equals sales volume multiplied by unit price, the growth in sales in value can be explained by the growth in sales volume and increase in unit price. This is shown in Table 2.

Table 2: Growth of Sales in Value, Number of Units Sold and Unit Price

	Time Period	Growth of Sales in Value	Sales Volume	Unit Price
GM	1951–1960	5.4%	1.6%	3.8%
	1961–1970	3.9 (8.0)	1.3 (6.4)	2.6 (1.5)
	1971–1980	11.9 (10.6)	3.0 (2.3)	8.7 (8.1)
Toyota	1951–1960	34.5	29.5	3.9
	1961–1970	22.2	26.4	–3.3
	1971–1980	15.2	7.5	7.3

Remarks: GM suffered some of its worst financial difficulties in 1970. If we were to calculate the average yearly growth in sales for 1960–69 and 1970–79, the rate would be 6.4% during the 1960s, 2.3% during the 1970s, showing substantial sales growth during the 1960s.

As you can see, the rise in GM's sales since the 1950s is attributable more to the increase of unit price than to the growth of sales volume. This was put forward by an individual named Donald Brown, who established a financial management system during the 1920s based on a policy of "target return pricing." During the 1950s, under the Sloan management system, a discrimination policy among products as well as a price management policy were strengthened. At the same time, with Ford and Chrysler, GM formed an oligopoly in the U.S. marketplace. Although sales did not increase much during this 10-year period, it was possible to increase prices at a level that exceeded the rate of inflation, and GM experienced an annual growth of sales in value of 5.4%.

Note: The Sloan principle is described in detail in the book *My Years with General Motors* by former chairman of General Motors Alfred Sloan, Jr. (Doubleday, 1963). According to Kiyoshi Yamazaki's book (mentioned earlier), this principle recommended pushing consumers to buy higher-end cars through the full-line policy, that is, the policy of introducing multiple models from Cadillac to Chevrolet.

During the 1960s, GM experienced a surge in sales as the American economy prospered. The increase in sales volume had a large effect on the growth of sales in value. On one hand, the price per unit during this time changed to a level comparable to the CPI. But, taking into account that the demand weight shifted from large cars to compact cars, the price increase was quite a large one. This enabled GM's pricing policy to take full effect.

During the 1970s, the auto market did not see much of an increase in sales, but there was a large price increase every year, in the magnitude of 8% to 9%, sometimes surpassing 10%. The rise in car prices during this period can be attributed partly to inflation due to the oil crisis and the resulting price rise in raw materials, as well as to cost increases to meet emission control and safety requirements. In this way, GM was able to gain a substantial increase in growth in sales in value under a downwardly rigid pricing policy.

The year 1980 saw a drop in growth of sales in value, a 13% drop from the previous year. This was caused by the sudden shift in market demand to small cars as a result of the second oil crisis, and an overall decrease in number of units sold due to the recession. This shift to small cars made it difficult for GM to maintain the managed pricing policy that had enabled it to operate under the existing oligopoly. Namely, there was a shift in unit price to lower-priced cars, and American, Japanese, and European products were all competing in the same area. For this reason, GM started programs such as the World Car and World Truck, planning to increase market share in its world markets.

In comparison, Toyota differed from GM in that it had a remarkably high growth as a result of its increase in sales volume.

During the 1950s, Toyota reinforced its production facilities under the government's protective policy. The company also benefited from the special procurements of its trucks by the U.S. Army for the use of the newly organized Japan Police Reserve Force (later, renamed as the Japan Self-Defense Force). Having enjoyed the overall economic prosperity at the time in Japan, Toyota increased its annual production on an average of 30% each year. Although its unit cost increase was almost at the same level as GM, its annual growth of sales in value increased 34.5% on the average, a figure that far exceeded GM's level of growth.

Note: The U.S. Forces placed special procurement orders for trucks with Japanese automakers one month after the outbreak of the Korean War on June 25, 1950. They bought from Toyota, Nissan Motor, and Isuzu Motors a total of 9,956 trucks (or 10,206 trucks, according to another report) worth $23 million. They also purchased auto parts and 1,908 units of engines, which accounted for 20% of the total value. They were delivered from October 1950 through May 1951 (according to *Automobile Section News No. 11* published by the Automobile Section of the Ministry of Commerce and Industry, July 1951). Toyota received orders twice for a total of 3,325 trucks and additional orders for 230 vehicles for the Japanese police reserve force (according to *The Thirty-Year History of Toyota Motor Sales*). The Korean War came to an end in July 1953. But the war in Vietnam broke out shortly thereafter. Then the U.S. Forces requisitioned their military-use trucks supplied to the Japanese Self-Defense Force and sent them to Vietnam. In return, the U.S. Forces procured Japanese-made vehicles and gave them to the Self-Defense Force. The trucks used in the Korean War had been produced based on each automaker's specifications, but in this case, the U.S. Forces looked for Japanese makers who were interested in producing trucks based on U.S. Forces specifications. Rigorous tests were conducted on all trial products at the U.S. Forces testing ground in the U.S. from the end of 1956 to 1957. Toyota, Nissan, Isuzu, and Mitsubishi participated, and all the firms passed the tests. The American Procurement Agency (APA) then called for bids for the trucks in 1958. Toyota made successful bids for all trucks in intense competition, and received orders for 9,364 trucks in 1959. All the trucks were equipped with diesel engines. Such special procurements were of considerable help to Toyota in achieving sound financial footing. And the U.S. Forces' strict guidance in production and quality control was also quite useful in Toyota's later development.

In the 1960s, spurred on by MITI's "People's Car" program, Toyota introduced its entry-level car, called the Publica. Not only did it set the stage for a motorization age supported by the high-growth economy, but Toyota was able to expand its sales network and make product diversification. This resulted in annual growth of 26.4%. Although its price per unit was decreasing, Toyota experienced remarkably high growth as a result of its increase in sales volume. This was brought by the severe price competition among Japanese domestic auto manufacturers and their efforts to reduce cost through mass volume production. In addition, demand was centered upon low-priced, mass-market cars such as Publica.

The 1970s saw a decrease in Japanese domestic demand due to the maturity of the car market and the shock caused by the first oil crisis. But there was rapid growth in overseas markets, especially the U.S., and as a result, Toyota's annual sales grew at a rate of 7.5%. In addition, there was an increase in cost due to the emissions control regulations, as well as increase in export prices. As for growth of sales in value, it increased at an annual rate of over 15%. In this high-growth environment, Toyota became number one in terms of sales value in 1975 among all Japanese industries, and ranked third among the world's automobile manufacturers.

Profit Margin (Ratio of Profit Margin to Sales)

Table 3 shows the ratio of profit margin of each decade for GM and Toyota. GM's margin during World War II was low (approximately 4% to 7%), because it was producing vehicles used in the war. With the recovery of civilian demand after the war, the company's production volume returned to the prewar level, and its profit margin improved gradually. By 1950, it increased to 11.1%.

Table 3: Ratio of Profit Margin to Sales

	GM	Toyota
1950	11.1	4.5
1960	7.5	7.4
1970	3.2	3.8
1980	–1.3	4.3
1990	–4.6	4.8

Under the Big Three oligopoly of the 1950s and 1960s, and as a result of the smoothing of target return pricing, which had maintained the ROE (return on equity) at 20%, GM's average annual profit margin on sales was 7.9% during the 1950s and 8.9% during the 1960s as the sales volume increased greatly. But with the 1970s came sudden increases in CPI and labor cost. During the latter half of the decade, measures were implemented to comply with government regulations, and greater investment was directed toward smaller cars. All of these contributed to the trend of a gradually decreasing profit margin. For this reason, the ratio of profit margin on sales dropped to 5.3% during the 1970s. As one can see from the way GM reacted to the strike that took place in the Lordstown plant and to the Vega defect problem, GM underwent a transformation from a more conservative management system. This happened in the midst of a surge in public criticism regarding management of the American auto industry. Consequently, GM was unable to move forward positively toward a strong management policy under the financial management leadership.

In 1980, because of the second oil crisis and the sudden shift to small cars, GM suffered a loss of $760 million and was in the red for first time in 60 years.

Note: According to Kiyoshi Yamazaki, GM's efforts to change its corporate culture included the establishment of the Public Policy Committee (1970), the appointment of African-Americans as directors (1971), the organization of Ecology-Active Staff (1971), and the formation of the GM Science Advisory Committee (1971).

Toyota had faced a major financial crisis in 1949 and had experienced huge financial losses. It was saved by financial aid offered by a consortium of 24 banks. As a condition for receiving this aid, Toyota was forced to split into two companies, sales and manufacturing. Its business performance quickly improved in 1950, when special procurement orders after the outbreak of the Korean War allowed it to expand production on a large scale, just as the company embarked on a reconstruction of its business by spinning off its sales division. It continued to increase its profits through efforts to promote efficient production, such as the introduction of the *kanban* system (the "Just-In-Time" production system) and improvement in production control. As a result, by the late 1950s Toyota was able to stand shoulder-to-shoulder with GM in terms of profit margin.

During the 1960s, as Japan's motorization advanced, price competition became severe, and Toyota's profit rate decreased from the levels of the previous decade. But with mass production and cost reductions, Toyota was able to increase and stabilize its profits.

In the 1970s, Toyota overcame a slump triggered by the first energy crisis by thoroughly streamlining its management, and maintained high and stable profits thereafter.

Toyota's high profit rates can be attributed to the following reasons. First, Toyota cut inventory as much as possible by introducing the *kanban* system to curtail inventory costs, and achieved cost reductions through its commitment to streamlining management. The separation of its sales department into an independent company helped considerably to reduce trade receivables and improve efficiency of fund operations. Toyota earned huge financial revenues without debt management. Incidentally, financial earnings for 1980 consisted of approximately 35% of total profits.

Liquidity (Current Ratio)

GM's liquidity in hand has been declining over the long run. As shown in Table 4, in 1950 and 1960 GM maintained a good fund position with a current ratio exceeding 100%. This favorable fund position was attributed to the firm's high profitability. In the 1970s, however, its liquidity in hand started fluctuating. The current ratio fell short of 100% in five out of the 10 years in the 1970s and set a record low of 65.7% in 1970 and 71.1% in 1974.

GM's fiscal policy was such that it procured its needed capital during the recession, a period of deteriorating profits, by selling securities rather than by resorting to long- and short-term loans. If sales were sluggish, earnings fell, resulting in the reduction of dividends as well as the sales of securities, or a decline in the current ratio. When sales revived, earnings rose, resulting in the purchase of securities or an improvement in the current ratio. This was repeated over and over again.

In 1980, the financing situation worsened further, and changes started to appear in GM's countermeasures. For example, sales of securities were replaced by long-term loans. From a self-financing-centered financial policy, which had been strongly adhered to, GM switched to long-term loans, thus to meet an enormous sum in capital investment, amounting to $80 billion.

Note: According to the January 1981 issue of *Ward's Auto World* and the January 26, 1981 issue of *Automotive News*, GM planned to invest a total of $80 billion over 10 years up to 1990. The automaker's investment plan from 1980 to 1984 consisted of a $32 billion project to reorganize its product lines, shift its focus to the front-engine front-wheel-drive (FF), and modernize its facilities in North America, and an $8 billion project to build new plants in Spain and Mexico. GM planned to invest another $40 billion in the downsizing of its cars, introduction of robots, construction of a new parts plant, and development of electric cars from 1985 to 1989.

Table 4: Current Ratio

	GM	Toyota
1950	179.9	36.8
1960	150.1	93.7
1970	65.7	109.7
1980	61.0	139.0*

*For 1979

Since Toyota had revived its finances in the early part of the 1950s through outside financing, short-term loans comprised approximately 40% of the liquidity debt, and its current ratio was extremely low in 1950, at 36.8%. However, as a result of the separation of the sales division, short-term liquidity financing was freed up, and by promoting a debt-free management policy, on-hand liquidity took a favorable turn. Thus, good liquidity results were evident from the mid-1950s.

During the 1960s, the current ratio improved to the 150% to 180% level, but in 1967, it decreased to 110% as a result of the increase in production of the hot-selling Corolla. This required an increase in raw materials, parts, and accounts payable. Afterwards, this current ratio remained unchanged. Beginning with the purchase of securities with surplus funds, like GM, Toyota established a pattern of monitoring its on-hand liquidity with the purchase and selling of securities. Even during the 1970s, its current ratio was stable at 110% to 130%, and Toyota completed repayment of its short-term loans during the latter half of the period. As a result, Toyota became completely debt-free, and its on-hand liquidity was in excellent condition.

Long-Term Stability (Rate of Owned Capital)

The ratio utilized here is calculated by the following formula:

$$\text{Rate of Owned Capital} = \frac{\text{Owned Capital}}{\text{Owned Capital + Long-Term Loans}} \times 100\%$$

Traditionally, GM used its own funds for capital investments and other fixed assets. This was based on the traditional tenet of American capitalism: "The company belongs to the shareholders, and with the exception of temporary needs for capital, there will be no reliance on outside financing." The costs of self-financing are higher than those of borrowing in the U.S., where stockholders require companies to pay high dividends. GM, which had always acknowledged itself to be the benchmark of American capitalism, had fully protected its earnings by way of its pricing policy. Furthermore, it had to adhere to its longstanding principle to share prosperity with its shareholders. This basic principle, however, was transformed drastically in 1980, thus bringing a major change to the 80-year tradition of the American auto industry.

GM did not have any long-term loans and had an owned capital ratio of 100%. But in 1954, a record sales year, required investment funds increased, exceeding the estimated level. Thus, it issued corporate bonds of $300 million at face value.

GM's U.S. operations rarely had to procure funds from outside sources until it issued $800 million of corporate bonds in 1972 to cope with the effect of strikes and the long-term declining profit. In 1975, the U.S. economy, battered by the oil shock of 1973, plunged into a recession. As inventory stockpiled, GM resorted to consumer rebates as incentives. As a result, unable to earn enough to cover its dividend pay-out and capital investments, it had to procure a long-term loan of $600 million. In 1980, a depression of unexpected magnitude hit the marketplace, and GM had to borrow as much as $1.3 billion worldwide, resulting in a decrease of its owned capital ratio to less than 90% (Table 5).

Note: According to its 1989 annual report, the U.S. unit of GM obtained a six-year loan on bills worth $200 million at 10% interest and an eight-year loan on bills worth $200 million at 12.2% interest. The automaker's overseas subsidiaries also took on new loans totaling $900 million, including a loan by the Canadian subsidiary at 18.3% interest. As a result, GM group loans reached a total of $1.3 billion.

Table 5: Long-Term Stability
(Rate of Owned Capital: %)

	GM	Toyota
1950	100.0	81.0
1960	94.9	81.1
1970	97.2	89.0
1980	89.6	100.0

Having lived through the management philosophy motto "One should mind his own castle," formulated during the 1949 crisis by then-president Taizo Ishida, Toyota eliminated debt to the best of its ability and established a policy to solidify its owned capital. In the 1950s, investments were continually made to renew equipment and facilities as part of the Five-Year Modernization Plan, and funds when needed were procured through loans from domestic and overseas banks.

From the end of the 1950s to the early part of the 1960s, Toyota made aggressive investments by increasing car production capacity or defending itself against liberalization of trade. Meanwhile, the balance on long-term loans continued an upward trend so that Toyota's owned capital ratio fell from the 1950 rate of 80.6% to 75.8% in 1965.

After 1966, as growth of sales in value increased and costs were reduced by mass production, greater accumulation of capital become possible. This led to the realization of an aggressive, fully owned capital policy, which was a countermeasure against capital liberalization, and as a result, bank loans suddenly decreased. When capital was needed, corporate bonds were issued. The amount of the new long-term loans did not differ much from what it had been during the first half of the 1960s, but corporate financial strength became very stable and secure.

Owned capital rate for 1970 was nearly 90%. In addition, according to the settlement of accounts issued at the end of the fiscal year in November 1972, when the annual production level had reached 2 million units, the new long term loans amounted to zero, thus marking the beginning of a debt-free company. In 1976, the repayment of the balance on long-term loans was completed, with the exception of one year, and the owned capital rate hit 100%.

Toyota attained a stable profit rate because of its management policy, which represented total rationalization or what is sometimes labeled "wringing a dry towel." Furthermore, it was able to strengthen its financial position and quickly develop a total owned capital policy as it headed toward the stable growth period of the 1980s.

Vitality (Rate of Total Assets Turnover)

GM's rate of total assets turnover around 1950 was about 2.0. It later dropped and remained at the level of 1.5 to 1.7, less than that of Toyota, until the middle of the 1970s. However, it started rising in the later half of 1970s, because GM's rate of sales growth exceeded that of total capital in use through surge in the price per car and sales (Table 6).

Table 6: Rate of Total Assets Turnover

	GM	Toyota
1950	2.40	1.28
1960	1.69	2.17
1970	1.29	2.07
1980	1.73	2.09*

*Represents rate in 1979

Toyota's figure was as low as 1.3 in 1950. In the following decade, however, total assets grew ninefold, and growth of sales in value was up to 13 times. As a result, the figure for turnover of total assets exceeded 2 in the late 1950s.

In the 1960s, total assets growth increased substantially, such that the turnover rate decreased to 1.5. But the latter half of the decade saw a widespread increase in growth of sales in value. As seen earlier, the turnover rate stabilized at 2 from 1967, revealing a vitality that exceeded that of GM.

Financial Strength of Owned Capital (Self-Financing Ratio)

Financial strength of owned capital is an indicator that measures how funds for capital investments in fixed assets, working capital investments, and dividends can be procured in a stable manner through owned capital. Whereas rate of owned capital is one indicator of the strength of owned capital as a static measure, financial strength of owned capital is a flow measure that indicates change from year to year (Table 7). The formula is as follows:

Financial strength of owned capital = (net profits + depreciation)/ (current capital investment + increase in working capital + increase in other long-term assets)

Table 7: Financial Strength of Owned Capital

	GM	Toyota
1950	100	NA
1960	94	52
1970	61	73
1980	38	71

GM spent its capital during the 1950s and 1960s in the following manner: 52% to 53% for investment in facilities, machinery, and tools; 38% for dividends; and 9% to 10% for other. During the 1970s, investment in facilities, machinery, and tools increased to over 60% to comply with various government regulations and to develop small cars; therefore, the percentage allocated for dividends dropped to below 30%. These funds were financed basically through owned capital. Most of the owned capital came from net profits and depreciation. Specifically, during the 1950s and 1960s, because GM had high earnings and many short-term capital investments with short depreciation periods, 95% of funds were financed through owned capital.

During the 1970s, earnings declined, and internal procurement became limited due to the increase in long-term capital investment during the latter half of the decade. But capital increases did not occur as much during this period. Perhaps this was because the cost of borrowing was lower than the cost of increasing capital, or because shareholders had voiced their disapproval of increasing capital. However, in 1979 and 1980, GM increased its capital to finance large-scale capital investments.

Toyota's financial strength of owned capital settled to 52% in 1960. The remaining amount of capital needed was financed through long-term loans, which comprised 19%, and also through capital increase, which was at 27%. The cash flow for 1970 was 73%, long-term loans constituted 2%, and others were 22% of total expenditures. However, most of "others" (21%) was due to an increase in specific reserve funds. In U.S. accounting terms, this is considered part of owned capital; therefore, to compare it to GM's figures, Toyota's financial strength of owned capital must be altered to 94%.

In 1980, Toyota's self-financing ratio was 71%, and reduction in working capital was 28%. Since the funds created by the compression of working capital were part of the owned capital that had accumulated in the past, its actual financial strength of owned capital was 99%. Thus Toyota was at the same level of financial strength as GM.

Long-Term Profitability (Return on Equity)

GM established a financial management system in the 1920s, targeting 20% return on equity (ROE). The company continued to achieve the target of 20% ROE on average until the mid-1960s, but the ROE began to decline in the late 1960s and dropped to 14% on average in the 1970s. In fact, there were times when the figure dropped to below 10%. In the midst of this downward trend, it was said that GM gave careful consideration to its enormous capital investment and extended its investment recovery plan from five years to seven years, reducing its ROE target to 15% (Table 8).

Table 8: Return of Equity

	GM	Toyota
1950	37.2	33.8
1960	17.1	27.3
1970	6.1	18.7
1980	−4.1	12.7

Toyota differed from GM in that it did not enforce a financial strategy based on ROE. But as a result of its good financial performance, Toyota was able to maintain a high rate of ROE. In 1950, because Toyota's owned capital was too small, its ROE was too high. Afterwards, as its owned capital was replenished, its ROE decreased. During the latter half of the 1950s, increase in earnings surpassed its owned capital increase, and Toyota was able to maintain an ROE at the 20% level.

During the first half of the 1960s, there were many occasions when capital increases occurred, and owned capital grew extensively while ROE decreased somewhat. Still, Toyota was able to recover to the 20% rate, increasing its earnings during the latter half of the decade. Even during the 1970s, Toyota was able to maintain a high level. However, after the first oil crisis, when business performance was down overall, Toyota's ROE was its lowest in 25 years at 11% to 12%. It recovered somewhat during the latter half of the 1970s, but as reflected by lowered earnings, Toyota was not able to maintain as high a rate as it had before.

Table 9: Difference Between GM and Toyota in Financial Structures

	GM	Toyota
Dividend ratio	70%*	16–18%**
Financial target	ROE = 20%	No-debt management

The ratio * was around 70% from 1945 to 1970. It dropped to 50% to 60% in the 1970s. The figures ** refer to the ratio from 1976 to 1980.

Table 9 clearly illustrates the difference in financial policy between the two auto manufacturers. The following formula is derived from GM's ROE equaling 20% and its dividend ratio equaling 70%.

ROE = 20%
= net profit per share/owned capital per share
= (dividend per share/owned capital per share) + (current reserve profit/owned capital per share)
= 14%(a) + 6%(b)

where (a) is the expected profits for stockholders, and (b) is internal corporate reserve and will be invested in fixed assets.

Since the market prices of GM stock were about 2.2 or 2.5 times the book value, the yield rate of 14% on the book value was equal to the yield rate of a little more than 6.4% on the market price. The actual yield rate of GM stock was 4.7% to 5.8% from 1966 to 1970, which far exceeded the industry's average rate of about 3%, and helped GM stay competitive. However, the adherence to the financial target of 20% ROE forced the company to recover invested capital in five years, and allowed it to avoid investing in machinery, which requires a long period to recoup, and focus on investment in jigs and tools, which can be recouped in as little as one to two years. This became the obstacle to GM's rationalization, as well as delaying development of new products. To give priority to protect shareholder profit, GM was obliged to sacrifice modernization of machinery and equipment and product improvements.

U.S. corporate management runs the company on behalf of stockholders and takes responsibility for the sound operation of the company — so-called trustee responsibility. They may be removed for breach of duty if they fail to pay dividends to stockholders. In contrast, Japanese companies need stockholders' approval for dividends, but the decision on dividends is essentially made by corporate management, based on their experience, and is approved at a general meeting of stockholders. Thus, Japanese companies generally pay out little in dividends. However, they do not come under much criticism, because Japanese companies share their stocks with each other, and low dividends help them cut their capital costs. In addition, Japanese stockholders do not always feel that they are the owners of a company. A famous social/legal scholar, Takenobu Kawashima, notes in his book *Legal Consciousness of the Japanese* (Iwanami Book Co.) that most Japanese shareholders have a notion that possession (management) takes precedence over ownership (stockholders).

U.S. management gives the highest priority to the improvement of profits for the stockholders, because they are watched by stockholders who are placing great hope in the future of the company. In contrast, Japanese management tends to give preference to the growth of the company, the improvement of profits, and the improvement of employee welfare over the interest of stockholders.

Productivity (Labor Productivity)

Labor productivity is defined by added value divided by the number of employees. The formula is as follows (see prologue):

Labor Productivity = added value/number of employees (value-added productivity) = added value/sales (value-added ratio) x sales in value/number of employees (sales per employee)

or

Labor Productivity = added value/sales (value-added ratio) x tangible fixed assets/employee (per capita tangible assets) x sales/tangible fixed assets (tangible fixed assets turnover rate)

Table 10: Labor Productivity

	Labor Productivity ($/Man)	Rate of Added Value	Per Capita Tangible Fiscal Assets ($/Man)	Turnover of Tangible Fixed Assets
GM				
1950	8,400	0.53	1,700	9.54
1960	13,300	0.62	4,900	4.36
1970	12,600	0.54	7,400	3.11
1980*	16,500	0.53	5,000	6.24
Toyota				
1950**	3,583	0.4	1,139	7.94
1960	8,639	0.25	5,972	5.70
1970	14,250	0.25	12,628	4.54
1980	30,167	0.22	19,626	6.92

Toyota's labor productivity and per capita tangible assets were based on annual average of U.S. dollar exchange rate (*IMF International Financial Statistical Year Book*, 1991). Figures of labor productivity and labor equipment ratio were deflated on a wholesale price index where the 1960 figures were equivalent to 100.

Remarks: *shows figures for 1979; **shows figures for 1953

GM's labor productivity was more than twice that of Toyota around 1950, as shown in Table 10. GM remained dominant in the first half of the 1970s but was surpassed by Toyota in the latter half, and the gap between the two companies has gradually widened. Toyota jumped ahead of GM in labor productivity, mainly because its per capita tangible assets (labor equipment ratio) far exceeded that of the U.S. company, while the value added ratio of GM, with a higher in-house production ratio, was 2.5 times that of Toyota. Until the first half of the 1970s, GM's labor productivity was higher than that of Toyota, mainly because of its higher value-added ratio.

In the mid-1970s, GM's rate of added value and equipment to labor ratio began declining, with the result that growth of its labor productivity began to slow down. This was attributed to GM's preference for short-term over long-term capital investment in the process of shifting its focus from its conventional large-car production to compact-car production. GM fell behind Toyota in terms of the labor equipment ratio and labor productivity. Around 1980, GM's per capita tangible assets were one-fourth those of Toyota, while its labor productivity was one-half, demonstrating a significant difference.

III-4. Conclusion

As mentioned in the previous section, GM gradually lost its vitality after 1950, when it was at the peak of its prosperity. The reasons for its decline are very clear. The following two points are of particular importance:

(1) GM was able to stick to its target return pricing with a target of 20% ROE over a period of years in the oligopolistic environment it had fostered. However, this management policy delayed modernization of its technology, products, and production facilities.

(2) GM gradually lost its competitiveness because it attached too much importance to stockholders' profits while ignoring corporate revitalization.

Toyota's management policy stands in stark contrast to that of GM:

(1) Toyota tried to use a snowballing method to expand its corporate operation cycle (discussed at the beginning of this chapter), giving preference to reinforcing its corporate strength, actively investing in product development and plant construction, allocating the biggest portion of its profits to operations and reinvestment.

(2) At the same time, Toyota adopted a management policy designed to correspond to changes in the market and satisfy the needs of the market quickly to the greatest extent possible.

As a result, Toyota succeeded in surpassing GM in terms of productivity and profitability in the 1970s. A tiny, insignificant Japanese company grew to be the world's largest automaker, equal to the Big Three within only 30 years. The source of its vitality was its struggle for survival.

Chapter IV

The Success of NUMMI

The Success of NUMMI

IV-1. Growing Popularity of Posts in NUMMI

New United Motor Manufacturing, Inc. (NUMMI) was set up in February 1984 as a joint venture in which GM and Toyota each hold a 50% stake. The new company took over the former GM plant in Fremont, California (which had halted production in 1982 due to an acrimonious dispute between the United Automobile Workers and management, and also because of its low productivity and inferior product quality). NUMMI began production at the end of 1984.

Toyota took the responsibility for operation of the plant. It sent 30 people to the joint venture, including the president (the first president of NUMMI was Tatsuro Toyoda, the immediate past president of Toyota), while GM transferred 17 people, including the executive vice president. Former GM workers who were UAW members were selected and re-employed by NUMMI. The joint venture reached full operation in the fall of 1986. It achieved quality and productivity rates that ranked comparably with those of Toyota's Takaoka Plant in Japan, which was rated the best automobile plant in the world.

Roger B. Smith, chairman of General Motors, said in a speech given at the Kansai Economic and Management Seminar in Kyoto on February 19, 1987:

> "We felt then, as we do today, that if New United Motor was successful, it would become a model for new American-Japanese industrial cooperation and would make a great contribution to both American and Japanese automobile industries. We felt that by creating this new international company we could both learn new approaches and techniques that would benefit both our companies and hopefully contribute to improving trade relations.
>
> "It has been four years, almost to the day, since Mr. Eiji Toyoda and I finalized our agreement. Today New United Motor is a reality. It has some 2,600 people working at the plant, and they have produced almost 300,000 cars since the first one was completed in December 1984.

"I should also point out that the majority of the work force at New United Motor is the same work force that was at the old General Motors Fremont plant until 1982, when the plant was permanently closed. Under this new system, most of these same workers, who were known for being radical and adversarial, are now the biggest supporters of the system. They are productive and, by far, the main reason for quality products being produced in Fremont.

"But, more importantly, when we look at Fremont, we see a different way to do business. And I believe I can say that both of us, GM and Toyota, have learned a great deal from studying the difference.

"At GM, we know that, for us to be competitive in the small-car market, we must change. We must try new approaches to ensure that we remain competitive and that we prevent further erosion of U.S. auto production and U.S. jobs. We are in Fremont to learn, and learn we have.

"Just as I know that Toyota will use the knowledge they have gained from the joint venture in their operations in Georgetown [where a plant of Toyota Motor Manufacturing U.S.A. (TMM), Toyota's wholly owned subsidiary, is located] and Cambridge [which has a plant of Toyota Motor Manufacturing Canada (TMMC), Toyota's wholly owned subsidiary], the lessons we have learned will be incorporated throughout GM. The knowledge we have gained in Fremont will be very important to the future of our company. And I think the same could be said for Toyota."

The book *The Machine that Changed the World*, written by Daniel T. Jones of Sussex University, James P. Womack and Daniel Roos of MIT (Rawson Associates, 1990) praises the Toyota production system, which is the best in the world. The authors point to the fact that "lean production," as they call it, could in its entirety be successfully transplanted to the U.S. from Japan despite the different environments.

Although GM's chairman and outside scholars have given NUMMI much praise, GM moved on without fully recognizing that it needed to introduce a new system. Not only that, but the experience and knowledge gained by those employees who had been transferred to NUMMI went unnoticed once they returned to GM.

That changed, however, beginning with the visits to NUMMI by president Lloyd E. Reuss and vice president William E. Hoglund in November 1990. These were followed by visits from vice presidents F. Alan Smith, Robert T. O'Connell, and John F. Smith (the present GM chairman) over the next several months. Furthermore, the majority of these top executives went to the plant before the 6:00 a.m. start of the workday.

At Toyota, designated work hours are strictly adhered to, and operations are at full speed during those hours. Preparations to begin or end work are not included in regular work hours. If the fist shift begins at 6:00 a.m., the workers are at their designated posts and work begins precisely at 6:00. This is not the case in American plants, where the start of the work day is delayed as laborers prepare themselves for work, and before the workday ends, they begin putting their tools away. The executives found this scene "exciting": workers in their place at exactly 6:00 a.m. and working. Since then, GM's directors and managers have continued to pay visits to NUMMI, and today two sets of teams visit daily to study various areas. (There is an agreement between NUMMI and GM that a maximum of only 18 visitors are allowed per day. Every day, that limit is met.)

Seventeen posts at NUMMI are filled by GM personnel. They did not want to take the posts at first. However, as the excellence of Toyota's production system was recognized both inside and outside the company, 200 or so GM employees applied for each job opening. It is said that as a result of this intense competition, highly qualified individuals were chosen.

Seven years after NUMMI began production, GM finally decided to study the Toyota production system as one of its corporate objectives. This had also been Roger Smith's goal, but it took seven years to realize it. The success of NUMMI was widely acknowledged.

IV-2. An Unrealized Partnership Between Toyota and Ford

The manufacturing joint venture between Toyota and GM was realized in the form of NUMMI, and it achieved great success in a short period of time. However, Toyota had initially tried to tie-up with Ford Motor Company, GM's biggest rival. When they failed to reach an agreement, Toyota started to negotiate with GM. Let's look at the course of negotiations between Toyota and Ford.

As a result of the oil crisis during the 1970s, which triggered the great sales increase of imported, fuel-efficient Japanese cars, the market share of Japanese imports shot up to over 20% of the American car market in 1979. The U.S. Congress, the government, and the UAW demanded that if this market situation were to continue, then Japanese automakers must also produce cars in the U.S. The Japanese government and the mass media also strongly urged the automakers to build plants in the U.S., warning that Japan could suffer repercussions such as the limitation of vehicle imports into the U.S.

For example, U.S. trade representative Reuben O'Donovan Askew warned in an interview with the *New York Times* (November 15, 1979) that unless Japanese automakers made a large investment in establishing production facilities in the U.S., Japan could possibly suffer some type of repercussion. Vice chairman Irving Bluestone of the UAW said in a *Sankei Shimbun* interview (November 28, 1979) that the U.S. would boycott Japanese cars unless Japanese automakers gave consideration to local production. Congressman Charles A. Vanik told the *Nihon Keizai Shimbun* (December 15, 1979):

> "If the Japanese automakers decide to build plants in the U.S., it will not only be a plus for the Japanese auto industry, but also for U.S.-Japan trade relations, and of course, it will benefit the American people. Without a doubt, everyone would agree that Japan produces high-quality products, and the idea that some of these Japanese products might be produced in the U.S. will receive much support from American suppliers and related businesses."

In January 1980, Honda Motor Co. announced that it would build a car assembly plant in Ohio next to its motorcycle manufacturing plant, which had begun operations in September 1979, and start production of passenger cars there in the fall of 1982. The announcement appeared to be a move to respond to public sentiment. Nissan Motor Corporation stated in January of the same year that it was preparing to build a pickup truck plant in the U.S. In October it selected Smyrna, Tennessee, as the site, and announced that it would begin operations in 1983.

In April 1980, assistant U.S. trade representative Robert David Hormats testified in a hearing before a special session of the House Committee on Small Business concerning the sudden increase in auto imports. He said that although earnest efforts were being made by Japanese automakers, such as Nissan's pickup truck plant and Toyota's investment in the U.S., more action would be needed. He called on Toyota specifically to act more constructively toward this end. Toyota had consigned research on investment in the U.S. to three research firms: Nomura Research Institute, Stanford Research, and Arthur D. Little (*Kyodo News Service* from Washington, D.C., April 21, 1980).

Ford Motor, having felt the full effect of the second oil crisis in 1979, plunged into a severe slump in sales and was forced to lay off many workers. It faced a large deficit that year and in 1980. On the other hand, Japanese automakers enjoyed an increase in sales of cars with better mileage. Eiji Toyoda, then chairman of Toyota, who had spent three weeks at Ford in 1950 to study its production system, had not forgotten the kind treatment he had received, and had always wanted to repay them. Believing that one should help out in time of need, Toyoda proposed to Ford the idea of joint production in the U.S. of a fuel-efficient compact car that Toyota was developing.

Toyota and Ford agreed to enter into negotiations on joint production in the U.S. A summit meeting was held at Toyota's head office when Donald Petersen, president of Ford, visited Japan in June 1980. However, the negotiations were somewhat rocky. Toyota's suggestion of producing in the U.S. a front-wheel-drive car, later named the Camry, was dropped because the car was likely to compete with the Escort/Lynx, a new compact car Ford had been developing.

Toyota then suggested another passenger car that was smaller than the Camry. But after the Ford delegation went back to the U.S., it became clear that this car would compete with another model for which Ford was promoting a development project. The negotiations were right back where they had started. Later, Ford showed interest in Toyota's TownAce Wagon and agreed to jointly produce a cab-over wagon developed from the TownAce in March 1981. However, when Ford conducted a product clinic on the prototype in California, they realized that the wagon was too small as a successor to Ford's Econoline.

The enthusiasm of the two companies for joint production eventually dissipated because the difficulty they faced in selecting a model (according to *The 50-Year History of Toyota Motor Corporation*). The companies announced that they had broken off negotiations in July 1981, one year after they had started.

In truth, it was the fourth series of negotiations between the two automakers on a joint venture had broken off since before World War II. Something always seemed to be standing in the way of the companies linking forces. As mentioned in the prologue of this book, Toyota and Ford opened negotiations for the first time in 1939 before the war. The subject of the first meeting was joint production of trucks by Toyota, Nissan and Ford. The three companies signed an agreement that Toyota and Nissan would each provide 30% of the investment and Ford would put up the remaining 40%. The plan fell through, however, due to political pressure exerted by the Japanese army, which insisted on domestic truck production.

The second series of negotiations were held in 1950. Toyota, which had had links with Ford before the war, expressed interest in a technical collaboration. Eiji Toyoda explained in his autobiography, *Toyota: Fifty Years in Motion* (Kodansha International, 1987):

> *Kamiya, president of Toyota Motor Sales, went to the United States first to negotiate with Ford. He agreed with Ford that the U.S. automaker would provide simple technological assistance to Toyota, and I was about to go to the United States to sign the contract. However, immediately before I left Japan, the Korean War broke out. Since the U.S. government then ordered engineers with important technological expertise to stay in the country, Ford was not able to send its engineers to Toyota as was specified in the contract. The negotiation had to restart from a blank paper. As it was not Ford's fault, Ford agreed to accept trainees from Toyota instead of sending its engineers to Japan. President Kamiya concluded a contract with Ford on the acceptance of trainees, and came back to Japan.*

As mentioned earlier, Eiji Toyoda had visited the United States in July 1950 to study plant operations at Ford factories.

The two automakers entered their third series of negotiations, concerning the joint production of the Publica, in 1960. They agreed on how much interest each company would hold in the joint venture, but the project was rejected by Ford's board of directors and did not come up again. The reason for the opposition still remains unclear.

As previously mentioned, the fourth series of negotiations, which lasted over a year, did not bear any fruit.

IV-3. Talks Quickly Shift to Toyota and GM

At the time, GM was being lauded as the number one automaker in the U.S. Having continued to downsize its large cars before the second oil crisis, it was able to overcome its problems relatively easily. However, GM was having problems with its small-car development program. Although the fuel-efficient, compact X-Car, which was introduced in April 1979, was declared a success at the beginning, major defects began to appear one by one. Large recalls were implemented over and over, and sales dropped rapidly. Furthermore, the subcompact J-Car, which had been introduced in 1981, also failed to meet early expectations.

According to *Rude Awakening* by Maryann Keller (William Morrow and Company, Inc., New York, 1989) when Roger Smith was appointed chairman in September 1980, he ordered that a cost study be conducted on the production of the "world car" or S-Car. The purpose of this study was to evaluate the production cost difference between GM's North American plants and Isuzu's. The results were as follows: The cost at Isuzu would be $2,857, whereas at a GM plant in North America it would be $5,731, a cost differential of $2,874. Upon these findings, Smith halted plans for the S-Car and decided to seek some type of cooperative venture with a Japanese automaker.

GM purchased 35% of Isuzu stock in 1971. Isuzu was known as a leading truck manufacturer, but it had a small presence in the car market. Smith asked president Toshio Okamoto of Isuzu whether GM could possibly buy Honda Motor, the third-largest Japanese automaker after Toyota and Nissan in Japan. According to Seiji Hirose, then senior managing director of Isuzu (later vice president of Isuzu and president of Jidosha Buhin Kogyo Co.), Okamoto did not approach Honda but told Smith that he himself believed Honda would never accept such a proposal.

Jay Chai, then vice president of C. Itoh, U.S.A., had previously served as the intermediary for the negotiations between GM and Isuzu. Aware of Smith's desire to seek a strong Japanese automaker as a partner, Chai immediately stepped in as an intermediary for GM and Toyota as soon as he learned that the Ford-Toyota discussions had been halted. In March 1982, Roger Smith and Eiji Toyoda met in New York and started discussions in earnest on the idea of a joint venture. Maryann Keller, the wife of Jay Chai, noted in her book *Rude Awakening* (William Morrow and Company, Inc., New York, 1989), that there was a rapport between the two executives that went beyond the usual business partner relationship, a sense of friendship characterized by respect and trust.

A few days after the meeting, GM and Toyota indicated that they were ready to start negotiations immediately to jointly develop and produce a small passenger car in the U.S. On February 17, 1983, Toyoda and Smith signed a contract for the joint production of compact cars at GM's Fremont plant.

It was not the first time GM had embarked on negotiations with a Japanese automaker about a joint venture. Genshichi Asahara, senior managing director (later president) of Nihon Sangyo (today's Nissan Motor), took orders from Gisuke Ayukawa, president of the company, and negotiated with GM from 1937 to 1938. GM, however, broke off the talks in 1938, saying, "It is not suitable to form a joint venture with a Japanese automaker in the present tense situation in East Asia. Adam Opel was requisitioned by the German government led by Adolf Hitler. The same thing is likely to happen in Japan." (A detailed account of the negotiations is contained in the record of Asahara's comments in *History of the Japanese Automobile Industry*, published by the Japan Automobile Industry Federation. I heard about this episode directly from him when I worked for MITI. He said that he had told Kiichiro Toyoda, later president of Toyota, everything about the negotiations between his company and GM.)

As I mentioned in an earlier chapter, Ford Motor had another stance for a joint venture with Japanese automakers. Ford management tended to think that politics and business should be separate, and signed an agreement to jointly produce trucks in Japan with Nissan and Toyota.

IV-4. A Strong Desire to Achieve a Joint Venture

The joint venture agreement was not readily accepted by GM's board of directors. President F. James McDonald, who objected to the project, said that he had carefully inspected Toyota's plants but had found nothing exceptional about the machinery and equipment and nothing to learn from Toyota. It is said that George P. Schultz, who was president of Bechtel Group and became an outside director of GM in July 1981, strongly supported the project, and that the joint venture was finally approved by GM's board of directors. Soon after this, Schultz became secretary of state in the Reagan administration.

Many at GM who had believed in the almighty GM were disturbed by chairman Smith's decision to halt the S-Car Project and go into a joint venture with a Japanese automaker to start production of subcompact cars under a Japanese production system. The general manager of the Research and Information Department at Toyota collected newspaper and magazine articles on the anti-joint-venture atmosphere at GM and circulated them in his routine overseas reports to Toyota's top executives, directors, and managers.

Upon seeing this report, chairman Eiji Toyoda became very upset. He said that it was inexcusable to spread this kind of information, which could jeopardize an important project — a project on which he had personally staked his reputation to make it succeed. He then ordered the information to be confiscated. Since Toyota was to be responsible for production, it was obvious that if this project were to fail, Toyota would have to take full responsibility. The success or failure of this venture would have a profound effect on Toyota's future. Eiji Toyoda's tenacity was very strong indeed.

A few members of Toyota's management were unsure about the venture. They wondered why they had to teach GM the Toyota production system, which Toyota had spent over 30 years developing and perfecting. But, moved by Eiji Toyoda's determination, they decided to exert all their efforts to make this project succeed.

Since Honda and Nissan had already made headway in the U.S., Toyota employees who had been waiting for Toyota to establish a plant in the U.S. had been preparing for that eventuality. Once the decision was made, all the employees were able to work quickly toward the same objective without detailed instructions or communicating across departments to get the job done.

According to Maryann Keller's *Rude Awakening* (William Morrow and Company, Inc., New York, 1989), the majority of GM executives refused to believe the production-related data regarding Toyota because it differed so much from GM's. However, Roger Smith believed the data. And he decided to go ahead with the talks—"Why not take the opportunity to get an insider's view of how the Japanese do what they do?" To calm the anti-Japanese and anti-joint-venture feeling within GM, Smith wrote an article, "Why the Japanese?" in the September 1982 issue of *GM Today*, an in-house newsletter. His main points were:

(1) Having the potential to supply components for Japanese cars from GM plants should have a positive effect on U.S. jobs.

(2) By abandoning the small car segment, GM could concentrate on larger cars, but this would be the wrong choice.

(3) Traditionally, smaller, less expensive models attract first-time buyers, and first cars can establish a lifetime of brand loyalty. If GM were to develop a small car on its own, it would take five years, too long a time to stay out of the small-car market.

(4) By marketing an entry-level car and attracting first-time buyers, GM would be protecting North American jobs both now and in the future.

(5) There was a great deal of "misinformation" regarding the negotiations between GM and various Japanese corporations. It was important that every GM employee understand why management was taking these steps and be assured that these negotiations were aimed at protecting the corporation and jobs in North America—at keeping GM competitive and productive.

(6) These steps should improve GM's worldwide position while providing it time to adjust efficiently and profitably to small cars and "reserve" a place in the market.

(7) The Japanese already had very efficient production facilities to produce cars like this in high volume. They also had the advantage of less government regulation. Combined with the undervalued yen, this meant they could ship these small cars at a much lower cost.

(8) Thus the strategy was to supplement GM's U.S. small-car output with some imported vehicles for the time being.

NUMMI's success was attributed to Eiji Toyoda's determination to make the joint venture a reality and Roger Smith's drive to overcome the opposition within GM. Kaneyoshi Kusunoki, who was in charge of the NUMMI project for Toyota, has said that without the friendship and trust that existed between these two top executives, there would be no NUMMI today. (Kusunoki, then an executive vice president of Toyota Motor Corporation, subsequently became chairman of TMS, TMM and TMMC, then chairman of Hino Motors. He was a protégé of Taiichi Ohno, who developed the Toyota Production System with Eiji Toyoda.) It wasn't rare for these two to pick up the phone to speak to each other when talks came to a standstill between the negotiating parties as they pushed forward to make NUMMI a reality.

IV-5. The Understanding and Cooperation of the UAW

NUMMI opened negotiations with the UAW, which had sought the reemployment of its work force, in May 1983, just after the joint venture project was unveiled. William Ussery, a labor consultant and the former Secretary of Labor under the Ford administration, served as the NUMMI negotiator, while the UAW was represented by president Owen Bieber and vice president Donald F. Ephlin. In the negotiations, these people abandoned the adversarial relationship that had existed before in labor-management relations and worked toward establishing a new agreement built on mutual trust. They realized that this was an integral part of the revitalization of the American auto industry. It was then decided that the labor agreement would begin with the statement that the relationship between labor and management would be based on mutual trust and understanding (labor-management relations in Japan and in Toyota are based on the same idea). Such relations were unprecedented in the U.S.

Note: The following statements can be found in the premise of the Collective Bargaining Agreement between NUMMI and the UAW and its affiliated Local Union 2244.

> *1-1. This Agreement is made and entered into this 1st day of July, 1985, by and between New United Motor Manufacturing, Inc. hereafter referred to as the COMPANY, and the International Union, United Automobile, Aerospace and Agricultural Implement Workers of America, UAW and its affiliated Local Union 2244, hereafter referred to as the UNION.*
>
> *The parties recognize that this is an historic endeavor and that progress for the COMPANY and the members of the UNION is to a large extent interdependent and therefore together we are committed to building and maintaining the most innovative and harmonious labor-management relationship in America.*
>
> *1-2. In the administration of this Agreement, and in our day-to-day relationship, we will exhibit mutual trust, understanding and sincerity, and to the fullest extent possible, will avoid confrontational tactics.*
>
> *Should differences or misunderstandings occur they will be resolved through full and open communication. The manufacturing environment will be based on team work, mutual trust and respect that gives recognition to the axiom that people are the most important resources of the COMPANY. We are cognizant that if this endeavor is to be a success, labor and management must work together as members of the same team.*

When Toyota started operation of its car assembly plant in Australia in 1974, it found that too many job classifications prevented the improvement of productivity, and asked the labor union to reduce them to five. As a result, productivity of the plant remarkably improved. At NUMMI, with the cooperation and understanding of the UAW, the number of job classifications was reduced to just two, consisting of operator and maintenance. The union also agreed to let instructors come from Japan to teach American workers the Toyota production system through nonverbal means — that is, by showing them how to do the work and through example.

One of the most important and unique characteristics of the Toyota production system is that it allows workers at the site to make job improvements on their own. The UAW said that job improvements should be made by engineers and not be the responsibility of labor. Of course, at Toyota, major improvements regarding equipment and work methods are the responsibility of engineers, but there are improvements that can be made only by the workers themselves. For example, if the corner of the check mechanism that grasps the workpiece is rounded slightly, it becomes easier for the check mechanism to grasp the workpiece, thus reducing error. This type of improvement can be made by the workers and can contribute greatly to increased productivity.

Toyota and the union agreed to call improvements that should be promoted by engineers "improvements" in English, as the union insisted, and those improvements to be made by the worker would be called *kaizen*, the Japanese word for "improvement." Today the Japanese words *kanban* ("Just-In-Time" production system) and *kaizen* are often found in English dictionaries, but perhaps it is only in this agreement that the fine distinction is made regarding *kaizen*. (This episode was related to me by executive vice president Kusunoki, one of Toyota's leading negotiators with the union.)

It was this understanding and cooperation on the part of the UAW that led to the successful launch of NUMMI.

IV-6. Universality of the Toyota Production System Is Proven

On September 22, 1983, Toyota, GM and the UAW concluded an interim management-labor agreement at the new plant. On April 11, 1984, the Federal Trade Commission (FTC) gave its approval for the joint venture between Toyota and GM to produce a compact car in the U.S. On December 11 of the same year, the first job at NUMMI came off the line.

The opening ceremony of the plant was held on April 4, 1985, and a three-year labor agreement between NUMMI and the UAW was approved by the UAW Fremont branch on June 25, 1985. In December of the same year, a second shift was added. NUMMI produced 64,601 units in 1985 and 205,124 units in 1986. Only two years had passed since the signing of the joint venture agreement when NUMMI started production, and a year later it had reached full production.

NUMMI held additional investment in its production facilities at a minimum. It introduced only a new press line and several robots. The establishment of this press line, however, was a very significant event in the history of the U.S. automobile industry. Until this time, American automakers had their main plants located around Detroit, and assembly plants that were located in faraway places such as California served merely as part of the automaker's distribution system rather than as the manufacturing network. It cost less to ship pressed panels and components from Detroit to California and build cars there than to ship completely assembled cars.

As will be discussed in detail in Chapter VII, U.S. automakers thought it was most efficient to produce as much as possible of one part at a time, because it took over 20 hours to do a press die change. Consequently, GM had never considered letting its California plant do press work. In contrast, Toyota was able to change press dies in only three minutes. The introduction of this Toyota technology enabled NUMMI to produce the required quantity of pressed parts at the time they were required, and at a lower cost than importing them from a plant in Japan.

The smooth production start at NUMMI proved that manufacturing technology developed by Toyota could be transferred to U.S. plants. NUMMI's success enabled assembly plants around the country to produce their own models, independent of major plants in Detroit. In this regard, the introduction of a press line to NUMMI's plant was of great importance. When I met with Ryoichi Kawai, president of Komatsu Ltd. (currently honorary chairman), which manufactures the presses, he told me that they had been receiving such large orders from GM, thanks to the success of Komatsu presses at NUMMI, that they were having to work hard to fill the orders. However, GM was not able to make most of the stamping lines they bought from Komatsu, according to *Comeback: The Fall & Rise of the American Automobile Industry* by Paul Ingrassia and Joseph B. White (Simon & Schuster, 1994).

The biggest challenge for NUMMI was the education and training of its workers. The majority of workers were former members of the most adversarial unit of the UAW, and their median age was over 40. NUMMI sent 300 people who were to become group or team leaders to Japan and trained them in the Toyota production system. Upon their return, they trained other workers.

At NUMMI there are no executive parking places, so the president must walk a distance if the parking lot is full. There is also no executive dining room; everyone from the president down to the line workers eats in one cafeteria. These changes were made to demonstrate in visible form the fundamental importance of respecting each individual in the Toyota system. The workers accepted this idea readily because it was the direct opposite of the former system, where workers were just a part of a machine and their abilities and ideas were ignored. In fact, they were not expected to have abilities or ideas.

The absenteeism rate at the time GM closed the Fremont plant in 1982 was 25%, but this figure decreased to just 3% at NUMMI. I visited the plant in August 1985 and was given a tour by president Tatsuro Toyoda. I was impressed by the pleasant atmosphere, where workers greeted the president as we walked by.

In 1986, not only was NUMMI named the number one plant of all the GM plants in terms of productivity and quality, but there was also talk that it was comparable to Toyota's plants in Japan. This became apparent when John Krafcik, who had worked at NUMMI as a quality control engineer (currently an employee at Ford), wrote a report called *Learning from NUMMI*, as part of an MIT study (to be mentioned in Chapter VII).

His report led to the publishing of a book, *The Machine that Changed the World*, by James P. Womack, Daniel T. Jones and Daniel Roos (Rawson Associates, New York, 1990). The book praised Toyota's outstanding production system, one that produced high-quality products in the most efficient manner, and further stated that this system could work with American workers and achieve the same results.

Roger Smith took every opportunity to call for the dissemination of the knowledge GM had acquired in its experience at NUMMI to other plants. A case where this actually happened was at the Van Nuys plant in California. Unfortunately, a lack of understanding between labor and management eventually led to the closure of the plant.

People from NUMMI had been dispatched to the Van Nuys plant to introduce the Japanese style of teamwork management. The Van Nuys branch of the UAW agreed that a production control system that would utilize the team concept would be introduced in May 1986. This was welcomed by the younger workers, but older workers with seniority were against the reform.

Bitter conflicts arose during the summer, and the production line stopped frequently. GM sent a letter to each of the 4,000 plant employees warning them that they were causing problems for customers and dealers. Subsequently, the union discontinued the team concept, and problems continued to plague the plant. In addition, sales of cars produced at Van Nuys slowed down, and the plant was finally closed in 1991.

As will be mentioned later, people from NUMMI were also sent to the state-of-the-art Saturn plant in Spring Hill, Tennessee, to establish a new production management system. However, many of the NUMMI alumni, upon their return to GM, were not given the authority or the positions that would enable them to implement what they had learned at NUMMI. Regarding this point, Kan Higashi, who was the second president of NUMMI and retired in May 1990 (currently president of Teleway Japan Corporation) said that this was something that he "regretted."

With the publication of *The Machine that Changed the World*, GM apparently realized for the first time the significance of the results at NUMMI and the importance of utilizing what they had learned. As I mentioned at the beginning of this chapter, GM's perspective on NUMMI has changed drastically.

In January 1992, Toyota and GM both agreed to extend the joint venture agreement, which would expire in 1996 if so requested by GM.

Chapter V

GM's Tough Fight

G M ' s   T o u g h   F i g h t

V-1. The Big Three's Late Comprehension of the Small-Car Market

In the early 1950s, when sales of European cars greatly increased, each of the Big Three began importing and marketing small cars manufactured by their own European subsidiaries. The Big Three did not work to develop these captive imports, nor did dealers try very hard to sell them since the profit on a low-priced small car was lower than that on a big American car. Ultimately the Big Three began manufacturing and selling compact cars in the fall of 1959 as a genuine countermeasure to the increasing imports of small cars.

These cars were acclaimed by Americans for their smaller size and lower price and were moderately well received by the market. However, compared to the leading European import, the Volkswagen, these cars were significantly larger and economically inferior in terms of gas mileage. And the Big Three had no inkling that there existed a large market for a subcompact car (this category was not so named until much later, in the 1970s), which was smaller than their smallest compact car. This was one reason they were unable to succeed in this category even in the 1980s.

Nearly all models of European cars imported into the U.S. market in the 1950s were also imported to Japan. I worked in the Ministry of International Trade and Industry (MITI) and was able to test-drive these cars. All of them were of good quality and performance; there was no striking difference among the various models. In spite of this, only Volkswagen showed significant growth in the U.S. market, because it established a new network of exclusive dealers as well as an extensive after-service network. If the Big Three had had the will, they should have been able to create an exclusive small-car sales channel that was at least no better or worse than that of imported cars such as Volkswagen and later Toyota. If sales of the captive imports of the Big Three had reached a certain level, those cars would then have been manufactured in the U.S. Thinking about it, I can only say this: The fact that the Big Three did not realize the change in the market and did not take proper action against the imported car market (subcompact car market) during the 1950s and 1960s is the main reason they face their problem today.

V-2. Corvair's Setback

In the late 1950s, when the import car share reached 10%, the Big Three began selling their own compact cars, which were smaller than their standard six-passenger large cars. The Ford Falcon and the Plymouth Valiant, downsized versions of their large counterparts, looked traditional. But the Chevrolet Corvair definitely looked different.

The Corvair had an aluminum air-cooled, horizontal, four-cylinder rear engine. It also had a beautiful, low-profile style. Mounting the engine in the rear allowed greater freedom in the design of the front body, but it was difficult to stabilize a rear-engine car. To build such a car was an adventure for the engineers. The Volkswagen, another rear-engine car, designed by the genius Dr. Ferdinand Porsche, was also vulnerable to cross-winds. (The Renault 4CV also had a rear engine, but had poor running stability when it was first developed. Not until after the end of the war, when Dr. Porsche was brought from Germany to help, did it finally go on the market.) Moreover, large-scale aluminum die-casting technology had not yet been developed, so the engineers had to overcome many obstacles before they could begin mass production.

Right after the Corvair first appeared, I drove one in California. But before I could fully experience its performance, its long fan belt broke and I got stuck on the freeway.

It was extremely well received when it first went on the market, but before long all kinds of problems occurred in succession. In 1965, attorney Ralph Nader published his book *Unsafe at Any Speed—The Designed-In Dangers of the American Automobile* (Grossman Publishers, Inc., New York, 1965) denouncing the defective design and performance of the Corvair. In addition, victims and families of victims who had suffered accidents with the Corvair filed lawsuits against the company. Although production was stopped in May 1969, annual production from 1967 amounted to no more than 20,000 units.

I am in no position to debate whether these problems were handled appropriately by GM, but I was impressed by the energy with which seemingly mature GM undertook this radical technological experiment. I suppose the reason they could embark on such an adventure was that they had sufficient resources to absorb the failure of any one car.

V-3. The Vega and the Death of the Ford-style Mass Production System

The U.S. small-car market expanded rapidly during the mid-'60s. Much of this growth was in subcompact imports. To counteract this situation, GM spent $100 million on the development of its first subcompact car, Vega, and built a new plant in Lordstown, Ohio, dedicated to Vega production. This new car was a traditional 2500-cc, front-engine and rear-wheel-drive model. The new plant, which aimed to be highly efficient, would utilize the ideal Ford-style mass production system. The goal was to produce 100 Vegas per hour without the use of physical labor. After production began at the end of 1969, 48,000 were produced in 1970 and 393,000 in 1971 — figures beyond what had been anticipated.

Buoyed by their success, in August 1971 GM announced that it was going to increase production levels to 140 units per hour for an annual output of 580,000 Vegas. But the union disagreed. A newspaper from that time (*Nihon Kogyo Shimbun*, January 27, 1972) reported as follows:

> *The laborers complained that the work flow speed was too fast, and they could not keep up with it. One car per forty seconds was too much to handle. They threatened to sabotage. To respond to labor's attack, the company resorted to severe measures such as closing down the assembly line and temporary dismissal of workers. As a result, labor and management relations continued to worsen.*

A recall was issued against 350,000 Vegas for defects in the throttle parts, and another 500,000 were recalled for a defect in the rear axle shaft. Vega production was continued until 1977, but volume could not surpass 1971 levels. It was the end of the Ford-style mass production era.

V-4. Chevette's Success

In September 1976, GM held a spectacular introduction for its new subcompact, the 1.6-liter-engine Chevette, which was to replace the Vega and compete against Japanese and European subcompacts. The Chevette had been produced in a plant in Brazil since 1972, and was the American version of the car that had been favorably received in Brazil. The same car was produced and sold in Germany as the Opel Kadett. It was also a best-seller there. Its gas mileage was EPA 25.5 mpg (rating by the Environmental Protection Agency) in city driving and 31.8 mpg on the highway, far better fuel economy than the Honda or the Volkswagen. As a result, the Chevette sold well from the very beginning. The following summer, GM decided to discontinue production of the Vega and to increase the daily production of the Chevette from 520 to 1,040 units.

Sustained by the increased price competitiveness against Japanese cars (due to a strong yen) and by the second oil crisis (which allowed high-priced but fuel-efficient Japanese cars to further increase their market share), the low-priced Chevette's sales were brisk.

GM decided to raise prices, citing an increase in production costs as a result of the oil crisis. These price increases were concentrated particularly on the Chevette, occurring four times from the beginning of 1978 through the end of July for a total increase of $400.

In August 1978, a GM spokesman made the following strong statement concerning the company's determination to battle the import tide: "If we look at this July's new-car sales, domestic cars grew 4.2%, while the imports took a big decrease of 7.7%. This declining trend, concentrated on Japanese cars, will continue from now on." Next, he talked about the favorable sales of the Chevette and said, "By introducing standard options and the four-door model, Chevette sales have grown 70%, and now it is the best-selling subcompact car in the U.S. To keep up with the strong demand, next February we will shift the production in one of our other plants to produce Chevettes." (*JIJI News*, New York, August 13, 1978)

At this time the exchange rate was fluctuating toward a weaker dollar, and imported cars, especially Japanese cars, had been forced to increase prices repeatedly since mid-1967. GM, which had long hoped for a decline in the price competitiveness of Japanese cars as a result of devaluation, instead took this golden opportunity to raise its own prices to maintain profits. Thus, it abandoned its efforts to preserve its competitive strength against the Japanese. This move was criticized as an opportunistic price increase even by U.S. business at large.

Around this time I had an opportunity to test-drive the Chevette in California. The style was called "Six Light" because, although it was a small car, there were three windows on each side. It was a nice-looking car. However, when I took it on the freeway, I was surprised by the slight vibration of the steering wheel. The recently completed freeway had tiny grooves cut into the road surface as a precaution against rain. But they affected the tires and caused the steering wheel to vibrate. Toyota some time before had discovered the same problem in the Corolla, immediately called forth the proper engineers, traced the problem to its source, and worked out a countermeasure. It would have been inconceivable for a Japanese manufacturer not to have taken some action more than a year after the car had gone on sale.

In spite of this kind of problem, the quality of the Chevette was not bad. In the *Consumer Reports* issue "The 1982 Cars," the Chevette was the only domestic car included in the "preferred car" list in the small-car group, along with imports such as the Honda Civic, Toyota Corolla, and Volkswagen Jetta. Sales of the Chevette continued at a healthy pace.

Then the EPA set a strict fuel consumption standard for each manufacturer, demanding a CAFE (Corporate Average Fuel Economy) of 27.5 mpg (22.6 km/liter) by 1985. In order to comply, the development of a front-wheel-drive car with excellent fuel economy became a must for each automaker. In 1978, GM announced the development of the J-Car, which would be "the new subcompact for the 1980s," to replace the Chevette. The J-Car was introduced in May 1981, but the production of the Chevette continued until the end of 1986. It established all-time production records for GM as a leading subcompact.

V-5. The Tragedy of the X-Car

From early on, GM engineers involved in the design of fuel-efficient cars had concluded that the best design was a front-wheel-drive car with the engine transversely mounted on the front axle in a right-turned position. Since the engine was started manually by a crank handle in early days, the direction of engine rotation was always clockwise. Even as we moved into the era when the engine was started electrically, it revolved clockwise. If a second axle were placed as a transaxle between the engine and the drive shaft, then the drive shaft (third axle) would revolve in the same direction as the engine. Consequently, the engine could be mounted transversely in a right-turned position, and if the third axle served as the driving force of the wheel, then the wheel would move in the direction of going forward. This design was said to be the simplest, and one that would minimize mechanical friction loss (this is based on a conversation with Masa-aki Noguchi, former senior managing director of Toyota).

Of course, early man knew that an object moved more smoothly if it were pulled forward rather than pushed from behind. In auto technology we also knew early on that front-wheel-drive would be the most rational design. A small car had the benefit of more interior room, so front-wheel-drive design was studied earlier in Europe, where small cars were being produced. However, it was difficult to design the car such that a change in movement direction could be managed at the same time that power was being driven into the front wheels. Furthermore, in those days, one had to design a front-wheel-drive car with the same engine as that of the FR car, a front-engine, rear-wheel-drive car, and in some cases the transmission also had to be shared. Until this time, no car in the world had this type of front-wheel-drive design.

It should be noted that GM reached the above conclusion after it had tested every front-wheel-drive car in the world.

In April 1979, GM's president, Elliot M. Estes, held a press conference in New York to introduce the X-Car, a front-wheel-drive compact with two types of engines: a four-cylinder 2.5 liter and a V6 2.8 liter. The fuel economy rating for this car met government standards; the 2.5-liter engine tested at 28 to 29 mpg (about 12 km/liter) in the city and the V6 at 25 mpg (about 10.5

km/liter). Estes said that GM had invested $2.7 billion in this program with the objective of surpassing the Japanese. Besides the Citation, which was to be sold by Chevrolet dealers, three cars from this platform with minor exterior styling changes were introduced on April 19 for the other three dealer channels: Pontiac Phoenix, Oldsmobile Omega, and Buick Skylark. They were off to an explosive start, setting a new sales record of 37,870 units for the first 30 days. In February 1980, the X-Car was named "1980 Car of the Year" by *Motor Trend* magazine (according to *American Automotive Chronicle, Research and Sources No. 33*, by Shoichi Inoue).

By August 30, 1980, production of the X-Car had surpassed the 1-million mark, a few months ahead of schedule. The 1-millionth car manufactured was the Chevrolet Citation, produced at the Tarrytown assembly plant in suburban New York.

But the X-Car faced repeated recalls due to defects; by January 1983 it had been recalled 11 times. It went down in U.S. automotive history as the car with the largest number of defects. When the brake defect problem arose in January, sales dropped 70% compared to the same month of the previous year. GM gave up on the X-Car, gradually reducing production until it was shut down completely at the end of 1985.

In spite of the $2.7 billion it had spent developing this car, GM lost it because of quality problems. GM's design concept had been very clear. Most of the front-wheel-drive cars that appeared throughout the world thereafter learned from GM's design and mounted their engines transversely in a right-turned position. Personally speaking, as an automotive engineer I felt deep regret that this car failed as a result of quality problems. The X-Car was a pioneer in the new era of front-wheel-drive cars.

V-6. The Collapse of the J-Car

GM earned record profits in 1977 and 1978, $3.3 billion and $3.5 billion respectively. Although it had an overwhelming market share in North America of 46%, in markets outside of North America sales of GM cars and trucks ranked fourth behind Toyota, Nissan, and Ford. In addition, although GM's profits in the North American market were twice those of Ford, Ford's profits overseas were more than twice those of GM. Buoyed by the favorable profit picture of the Chevette in Brazil, Germany, and the U.S., it is easy to understand why chairman Thomas Murphy and president Elliott Estes at the time believed in the "world car" concept.

In October 1979, Estes made the following points at the U.S. Foreign Trade Conference held in New York (*AP-DJ*, New York, October 22, 1979):

> "The world car will be a passenger car which will be produced based on a fundamental design — a design that will be universal in several countries. Therefore, we will try to utilize as many common parts as possible. With this system we would be able to meet the demands of each market by making minimum adjustments, and thus have competitive strength in most of the key markets. This type of integrated, global operation enables us to economize many of our costs.
>
> "One of the obstacles facing this world car is that those countries which have small-scale markets will have an obligation to establish a high rate of domestic auto production. The motive that obligates this type of domestic production can be understood, but this inevitably leads to extremely high costs and inefficiency.
>
> "One viewpoint concerning this obligation to produce domestically is as follows: An automaker from each of those countries can produce enough volume of one or two parts of the world car to minimize costs. Furthermore, each country can sell small volumes domestically and export the rest. As a result, each country will gain more confidence in exporting, and it will eventually become possible to import completely assembled vehicles."

The J-Car was designed according to this concept at the GM Technical Center in the outskirts of Detroit. The development project was also integrated with GM's West German subsidiary, Opel. Production was scheduled to take place in five subsidiary companies in five countries: Opel in Germany, Vauxhall in England, GM Holden in Australia, Isuzu in Japan, and GM Brazil in Brazil.

On May 21, 1981, sale of the J-Car began in America. It was an 1800-cc front-wheel-drive car with fuel economy comparable to that of Japanese cars at 29 mpg city (about 12.4 km/liter) and 45 mpg highway (about 19 km/liter). Besides the Cavalier, which was sold by Chevrolet dealers, the Pontiac J2000 and the Cadillac Cimarron were introduced at the same time. President F. James McDonald proclaimed the J-Car to be the best that GM had ever produced, excelling in fuel economy, price, and quality. He further emphasized that GM had invested a record $5 billion in its development. Three months prior to this event, McDonald made the following statement at a press conference in Cleveland (*AP-DJ, Kyodo Newswire*, Washington, March 19, 1981):

> "GM will stand face-to-face with the competition from Japan. We will not pull out from the competition with Japanese cars. We intend to fight head-on with anyone, be it West Germany or Japan. And we will prove to the American people that price and quality are not words in a foreign language. GM's objective with the J-Car is to add a punch to the muzzle of the Japanese. The J-Car will be a good match for similar-looking Japanese cars."

But, in reality, the true price of J-Car fell somewhere between that of the Chevette and the Citation, much higher than the expected price range. In fact, its price was closer to that of the upscale X-Car, the Citation. Customers coming into the showroom compared sticker prices and experienced what the news media called "sticker shock." Consequently, sales for the two months after introduction totaled only 32,000 units, far less than the approximately 173,000 X-Cars sold during the same time period.

Another problem had to do with engine performance. In August 1981, I had the opportunity to rent a Cavalier in Jackson, Wyoming, to visit Yellowstone National Park. It was an automatic, and I was surprised by its lack of power. It didn't even have the acceleration of a golf cart. I couldn't imagine how a company such as GM could build such a car. If GM's top executives had test-driven other competitive cars such as the Corolla during the development stage, they would have realized immediately that this car was far from a true product. I thought the reason this car was so under-powered was that GM had let its engineers in Europe design it. European requirements at the time differed in several technical areas from those in the U.S. First of all, emission regulations, which inevitably affect performance standards, were lax in Europe, and an automatic transmission, which reduces transmission efficiency, was not yet readily available in Europe. Second, since air conditioners, which consume much power, were rarely used in Europe, perhaps the engineers underestimated the amount of power that would be needed for cars sold to the American market.

However, I learned that my conjecture was mistaken. According to the book *The Decline and Fall of the American Automobile Industry* by Brock Yates (Empire Books, New York), Opel engineers had opted for the high-speed engine, but Detroit engineers insisted that a lower-power engine was enough. They claimed that American drivers didn't rev up their engines and therefore didn't need high-speed engines. Ironically, although they were advocating the merits of the world car, at the same time GM was at odds with the most essential part of the car, the engine design. That is, at the design stage, there was a difference in opinion on the J-Car engine to be produced in the U.S. and the one to be produced in Japan and Europe.

But this was no excuse for the lack of awareness on the part of GM's top management regarding rival imports. For example, in January 1981, prior to becoming president, McDonald said, "The J-Car ought to first strike a blow to the Toyota Celica." (*UPI-ES*, Detroit, January 2, 1981). If he had said "Corolla," it would have proven that he knew a little about imports.

In its January 1982 issue, *Consumer Reports* said in reference to the J-Car, "They should not be bought yet, and if anything, they are not at all suitable." The situation was much more serious than that. Since December 1981, a half-year after the introduction of the J-Car, production plants had been curtailing operations and stopping the lines. The great concept of the world car disintegrated and failed in its first act, the American marketplace.

V-7. The Rejection of the S-Car

Prior to the development of the J-Car, as part of the world car concept, GM was involved in designing a smaller mini-car, the S-Car.

Mini-cars were very popular in Japan, where 1.1 million units were produced in 1980, comprising 10% of total auto production. Besides consuming less fuel and having a low price and light sales tax, mini-cars had other advantages. It was not necessary to register them with the government, and it was relatively easy to obtain a driver's license for this type of car. Five Japanese manufacturers built mini-cars: Honda, Suzuki, Daihatsu, Mitsubishi, and Fuji Heavy Industries. All of them were manufacturers of two- or three-wheeled vehicles.

Although legal exceptions were made for the driving and registration of mini-cars, they were restricted in terms of vehicle size and engine displacement. Because of these restrictions, they were inadequate for driving on American roads. An attempt was made to export one with a much larger engine to the U.S., but its volume was limited.

However, the mini-car's fuel economy, rated at 50 mpg (21.2 km/liter), was outstanding. Perhaps as a result of the fuel crisis, which led to a greater emphasis on fuel economy, GM reexamined these superb fuel-efficient cars and decided to develop its own. One of their motives for attempting a buyout or joint venture with Honda and Daihatsu before chairman Roger Smith began his talks with Toyota about a joint venture may well have been their interest in the mini-car.

The S-Car, a two-passenger commuter car, was to begin production in 1985. Two types of engines, a three- and four-cylinder, were being considered. The former was a 1000- to 1200-cc engine. A Japanese machine tool manufacturer had received orders to prepare for production of a three-cylinder engine.

However, *Automotive News* (January 18, 1982) reported that Michael F. Juras, the Chevrolet chief engineer in charge of Small Vehicle Programs and Advanced Engineering Activities, said, "The S-Car ran into some financial difficulties, so it is not really a 'go' project by any stretch of the imagination." As a result, the order to the Japanese machine tool maker was canceled (*Nihon*

Kogyo Shimbun, February 17, 1982). This action demonstrated that the plan for domestic production of the S-Car had been abandoned. (Since there was no official announcement or statement regarding the S-Car project until this time, of course, there was no official announcement regarding its cancellation.)

Roger Smith based his decision to cancel the S-Car project on a survey that showed that it would cost significantly more to produce a mini-car in the U.S. than in Japan (see Chapter IV). However, according to Maryann Keller's *Rude Awakening* (William Morrow and Company, Inc., New York, 1989) the cancellation of the S-Car project was painful for Roger Smith. GM had to admit that this project would not be cost-effective and therefore had to be abandoned. Not only that, but the decision to jointly produce cars with a Japanese manufacturer demonstrated GM's limited capabilities. There seemed to be many at GM — more than before — who were too proud to cancel the program. It was Smith's decision to overcome their opposition.

In October 1981, GM acquired 5.3% of Japan's largest mini-car manufacturer, Suzuki, entering a joint capital relationship. In March 1982, president Osamu Suzuki had talks with chairman Smith in Detroit, and they announced that Suzuki would put together a design concept for the mini-car for GM by the summer and would supply the cars to GM starting with the 1986 model. In May 1982, it was announced that Isuzu, of whose stock GM now owned 37.5%, would supply 200,000 subcompacts of their own development to GM beginning in 1984.

In the May 26, 1982 issue of the *Wall Street Journal*, John B. Schnapp (at the time a vice president at Harbridge, a Boston-based consulting firm) wrote a short article called "GM Shakes Up the Auto Industry." After commenting on Michael Juras's statement about the cancellation of the S-Car, Schnapp went on:

> "This week the news came that GM in 1984 will start to import up to 200,000 subcompacts a year from Isuzu Motor Ltd. of Tokyo. This development confirms GM's apparent belief that from a long-term standpoint, 40% of the American passenger car market will probably be conceded to foreign automakers...But the future of the American suppliers of raw materials and components to Detroit will certainly be a diminished one. It appears unlikely that an industry — automakers,

material suppliers, components vendors — which used to produce in excess of nine million cars per year will ever again see a seven-million-car year."

In October 1986, GM and Suzuki established a joint venture called CAMI (Canada Automotive Manufacturing Inc.) in Ontario, Canada, with each putting up half the capital. They began joint production of mini-cars in 1989. Meanwhile, the European S-Car was designed and developed under Opel. The engine and transmission were manufactured at GM Austria and final assembly done at GM Spain. It had been sold under the name Corsa since 1982. By the next year, production had reached 20,000 units a month and surpassed 30,000 by 1988. Manufacturing and sales were limited to within Europe.

The new model of the Corsa was introduced at the Geneva show in March 1993. It had been designed by Hideo Kodama, a veteran designer who had joined GM immediately after graduating from a school of art and design in Japan, and worked there for more than 20 years. This new model received acclaim both in and outside Europe, and preparations are now being made to manufacture it in various locations around the world.

The success of the new Corsa was discussed in *Automotive News* (June 27, 1994) in an article entitled "The Accidental World Car":

Since 1982 the manufacture of the old 3- and 5-door models has occurred at the Zaragosa plant in Spain, 3-doors and hatchbacks at the Eisenach plant in the former East Germany, and the commercial and panel vans that form the basis of the Corsa at the Alzambugia plant in Portugal. Outside of Europe, production of the 3-door began at the Sao Paulo, Brazil, plant in March 1994. Production is also scheduled to begin at the Ramos Arizpe plant in Mexico in 1995.

Moreover, discussions are under way for a joint venture with Russia's largest automobile manufacturer, VAZ. It is under negotiation to produce it in China in 1996. Negotiations are also under way to have assembly of the CKD at Delta Motors in South Africa. By a twist of fate, after more than ten years the original goal of the S-Car as a world car has finally been attained with the second generation. The second-generation Corsa was intended for sales and production in Europe like the first one, but during the development stage, demand came from other markets and it began to be developed as a world car.

In the fall of 1993, I was visited by the planning director, general manager, and two other engineers from Ford. One of them was the man responsible for the development of the new Mustang, O. J. Coletti, whose venture is described in a chapter of *Comeback* by Paul Ingrassia and Joseph B. White (Simon & Schuster, October 1994), in which he is portrayed as the new hero of the American automotive industry. They asked about the key to the success of the Corolla as a world car. I answered, "The Corolla was developed as a car for the masses, and has the most customers of any car in Japan. This car was also well received in foreign markets, and only became the biggest volume car in the world by chance." GM's new Corsa backs up what I said.

V-8. The Vision of Saturn

In November 1983, chairman Smith of GM announced a plan for a small car called the Saturn that would compete against Japanese cars during the latter half of the 1980s. Coincidentally, around this time, the GM negotiation team, headed by vice president Jack Smith, was approaching the final stages of a joint venture on NUMMI at Toyota headquarters in Japan. When Smith learned about the Saturn, he couldn't help but express his shock because the subcompact production project was almost the same as the one he was attempting to manufacture at NUMMI.

After he was appointed chairman, there was one definite path that Smith aimed for: that GM must inevitably seek the cooperation of a Japanese automaker if it wanted to build a subcompact or a mini-car, because GM could not compete in that segment with the products from Japan. Because of this conclusion, GM canceled its S-Car project, and instead sought various means such as purchasing Isuzu stock, going into a joint venture with Toyota, and joint development work and production with Suzuki. Of course, there were many at GM who were worried about their company's situation. After all, GM was the number one automaker in the world and also represented American industry. It was disconcerting to think that now this company had to rely on the strength of Japanese automakers. In its November 3, 1983 issue, the

Wall Street Journal reported that "the announcement of Saturn's plans was a strategy on the part of GM to avoid criticism of GM as having 'given up its plans to develop a small car on its own.'" According to *Call Me Roger* by Albert Lee (Contemporary Books, 1988), the start of this project was based on the following ideas:

> *It started with three fundamental rules. First, was to go all the way— employ the most advanced technologies that could be found anywhere in the world. Second, Saturn was to be entirely new. There would be no carryover parts from other cars. Except for a few nuts, bolts and bearings, every part of the Saturn car would be completely new. Nor would any existing machinery or process be employed or old building used. This would be a company that was new in every way, one that used the most advanced concepts to rethink and reinvest every aspect of the automobile. Third, it would be entirely American. Not a single foreign part would be incorporated into the car; not a single foreign worker engaged in manufacturing.*

These concepts were discussed by GM's best and brightest engineers and managers, who had been selected from within. It was labeled the GM80 project and had started in June 1982, but it was also called the "No Year Project" because the plan didn't state when production would begin. Having abandoned the S-Car project, perhaps Smith, who had pushed for joint ventures with Japanese auto manufacturers by persuading employees who were against this idea, was trying to compensate for this debt by having the Saturn project. And once that was decided, he wanted to begin production while he was still in office. But Smith had bigger dreams. *Call Me Roger* describes the following:

> *Roger began writing his own scenario on the Saturn...He drew a picture of Saturn as ultra-high-tech. The buyer would sit down with the salesman at a computer terminal. He would say, "I don't smoke and don't want others smoking in my car," so the salesman would punch in "no ashtrays." He'd say, "I like the car to feel firm in curves and yet give a soft ride on the open highway." The salesman would punch in a tailor-made suspension system...While all of this was going on, the computer would check the buyer's credit...When he pressed the execute key, every one of the fifteen thousand parts that would go into his car would be automatically ordered from the right suppliers and computer-directed to*

arrive at the factory at precisely the right time for assembly. A robot would meet the shipment at the dock and deliver it to the high-tech assembly line. Highly flexible automation meant that a wide range of cars, with an infinite number of personal choice variations, would move down the same line...All of this would be accomplished—from buyer's first choice to delivery—in just two weeks, without a single piece of paper changing hands along the way. Orders, suppliers' invoices, billings, and even the monthly payments would be handled electronically. This entire scenario was pieced together from a wide range of interviews and speeches where Roger spoke at length about his out-of-this-world project.

Note: Roger Smith's dream was actually realized at Toyota in 1970 with the small specialty car, Celica, which was introduced in the Japanese domestic market in December. Customers could order the car with their desired options according to Toyota's newly implemented "Full Choice System" or "Car-puter System" (see Chapter I). This order system was similar to the one the chairman dreamed of.

In reality, the plants for this project did utilize some of the concepts learned at NUMMI. For example, after negotiating with the UAW, job classifications at Saturn were reduced to five, and the team concept was implemented. The executive parking and dining room facilities were also eliminated. At the same time, they attempted to utilize the "Just-In-Time" system and have "design-in" programs (see section VI-8) with their suppliers. GM's initial plan was to invest $5 billion and produce 500,000 subcompacts annually, but this plan quickly changed. First they decided to change the car from a subcompact to a compact because of intense competition in the subcompact market, and decreased their annual production figure to 250,000. Second, they cut their investment to $3.5 billion, having failed in their ultra-high-tech project at the Hamtramck plant (see next section).

The production of Saturn began in the summer of 1990, but by no means was it off to a smooth start. The production of the door, which utilized a plastic panel, became a bottleneck. Problems also arose in the production of the engine and the automatic transmission. GM defended the production slowdown by repeatedly claiming that they were adhering to the most important principle of producing quality products, and defective products would never be allowed to come off the line. Accordingly, Saturn's quality was much improved compared to GM's other products. Moreover, design problems such as engine noise or vibration were immediately brought to specialists for improvement. In the first year and a half after production began, machinery was barely operating at half capacity, and not until 1992 did production surpass 200,000 units.

The body of the Saturn is constructed of a steel cage equipped with plastic panels. Assembly was not done on a conveyor belt but rather by the skillet method, in which the workers and the car being assembled sit on the same moving "skillet," the workers moving around as they work. For this reason productivity was low, supposedly half that of Japan. Therefore, the only way to increase production was to work long hours, so three teams of workers alternated, two teams working 10-hour shifts each day. If the plant ran 20 hours a day, six days a week, an annual production of 320,000 units would be possible.

Sales were quite good until the summer of 1993, with demand outstripping supply. That autumn the company tried to force long hours to meet the fiscal balancing of revenues and expenditures for 1993. Each shift was extended 30 minutes so that the plant was running 21 hours a day, and a Sunday shift of 10 hours was added. In this manner a monthly output of 25,000 to 30,000 cars was reached, but ironically demand suddenly dropped to less than 20,000 cars a month. By the end of January 1994 there was a 100-day overstock, and a large-scale adjustment of production became necessary. As 1994 progressed, the addition of new retailers to the initial 221 outlets helped bring about a better financial situation, and sales gradually rose 25% over the previous year for a total of 286,000 cars.

Saturn carefully selected dealers who would always put the customer first. Moreover, an investment of $3 million was required of each dealer, so that a sales network of character and quality was established. For this reason customer satisfaction levels were the best by far of any dealership in America. At present the Saturn is available in sedan, coupe, and wagon models, but to continue its healthy sales trends over the long-term an additional model, preferably a luxury car, is necessary. It is said that recently there was a feasibility study on a plan to manufacture the Opel at another plant like the Saturn Spring Hill factory. Moreover, it has already been more than 10 years since

Smith announced the Saturn program and the time for a model change had come, so it will be interesting to see how GM will respond.

V-9. Hamtramck Plant Versus Tahara Plant

At the end of the 1970s, GM decided to replace an aged Cadillac plant in Detroit with a new car plant built in the same area, called the Hamtramck plant. Coincidentally, Toyota was also building a new car plant in Tahara, Aichi Prefecture, Japan, around the same time. The executives responsible for the construction of each of these plants, executive vice president Kaneyoshi Kusunoki of Toyota (also chairman of TMM at the time) and vice president Robert Stempel of GM (later president and chairman of GM), promised each other that they would meet once their respective plants had been built.

The Hamtramck plant had 260 robots, 50 automated guided vehicles, and a body shop where robots did 91% of all welding. It was definitely to be a showcase of high technology.

> *The assembly line passes through a perception unit, a bank of 121 cameras that measures every body opening — doors, windows, hood, and so on — in every car to within a half-millimeter's tolerance. The machine feeds the information back to a computer that automatically adjusts the welding robots, ensuring that all parts fit exactly.* (*Fortune*, July 8, 1985)

The Tahara plant greatly utilized transfer presses in its press shop, while in the body shop, three kinds of car models could be produced under a work flow line system with 87 welding robots. Drying furnaces were concentrated in the paint shop, and the assembly shop had slot conveyors and tilt hangers that could move the body up, down, or at an angle. Special attention was paid to the workers' postures and working conditions. In February 1981, the Soarer came off the line as planned, and production commenced in June for the new Celica XX and Celica Liftback, and for the Supra in August.

Completion of the Hamtramck factory had been planned for early 1983, but production did not begin until more than two years later, in August 1985. Working conditions, however, were absolutely terrible. The *Wall Street Journal* (May 13, 1985) ran the following article under the headline "Tricky Technology: Automakers Discover 'Factory of the Future' Is a Headache Just Now":

The Hamtramck plant, instead of a showcase, looks more like a basket case. Though the plant has been open for seven months, the automated guided vehicles are sitting idle while technicians try to debug the software that controls their movements. In the ultra-modern paint shop, the robots at times have spray-painted each other instead of cars. Some cars have been painted so badly that GM had to ship them to a 57-year-old plant to be repainted. Hamtramck is turning out only 30 to 35 cars an hour, far less than the 60 an hour it was designed to build.

Toyota chairman Eiji Toyoda had toured the Hamtramck plant on December 17, 1985, during his visit to Detroit, upon being advised to do so by Roger Smith. After the tour, Smith asked Toyoda for his impressions. Toyoda could only answer, "This is certainly a splendid plant."

In General Motors' annual report for 1985, magnificent color photographs showed an overall view of the Hamtramck plant and other high-tech production facilities. The caption reads as follows:

The first production cars came off the assembly line in September 1985. Approximately 21 miles of conveyors move vehicles throughout the plant. At full production 1,700 cars are in varying stages of assembly...Several loosely assembled subassemblies are precisely aligned and welded in less than 50 seconds by 10 electric robots in a single machine called robogates...A robot applies a bead of urethane sealer to the edge of a windshield glass before it is installed on a vehicle.

Reading this report, anyone would guess that the Hamtramck plant was Roger Smith's pride and joy. Apparently no one had guessed yet that this "ultra-modern plant, an incarnation of high technology" would turn into chaos.

The Tahara plant had gone to work on schedule. Upon completion of the plant, vice president Robert Stempel was invited by Kusunoki to visit Tahara. However, right before he was to leave for Japan, he was told by Smith not to leave Detroit because the Hamtramck plant was in such a precarious condition. Kusunoki later commented on this situation:

> "It is very difficult to upgrade production technology. Even at Toyota, we have wasted perhaps one-third of our investment in upgrading. The Hamtramck plant is said to be a showcase of the world's most revolutionary production technology. It must be extremely difficult to fix so many new experiments at the production site at once. We are paying attention to their results. If there was something that we would like the American engineers to learn from us, it is that they must go to the plant floor more often to see the operation, listen carefully to what is said, think about how to implement new technology. We have also told engineers at TMM [at which Kusunoki had formerly served as chairman] that they must spend at least one hour every day walking around the shop floor, or walk to the job site rather than taking a tram. In this way, they can only increase the opportunity to gain knowledge about the plant."

Toyota's luxury nameplate, the Lexus, was introduced in 1989. Comparing it to existing imports like Mercedes-Benz, BMW, and Jaguar, which had dominated the import luxury market, customers soon noticed that the Lexus was of much higher quality than the others. As a result, it surpassed the European imports in sales. The high-quality Lexus is produced at Tahara.

It can be said that with the establishment of these two plants, the destinies of these two companies had been decided.

V-10. The Key Reasons for Failure

GM's efforts to develop a small car during the past thirty years have not yet come to fruition. The following reasons can be pointed to:

1) Underestimating the imported small-car market

As sales of imported small cars took a sudden increase during the 1950s, some Americans began to question the efficiency of large cars — such as those made out of a two-ton lump of steel and a 400-horsepower engine — and instead expressed their desire for smaller cars. GM manufactured small cars in Europe. The sales of those cars in the U.S. contributed to the postwar revival and prosperity of Europe, then desperately in need of U.S. dollars.

As I discussed in Chapter IV, it was unfortunate that GM did not invest in establishing a separate franchise for its import cars at this time. GM ignored the signs of change in the tastes of the American consumer when small cars became popular in the U.S., and failed to capitalize on them.

During the course of the 1960s and 1970s, when sales of imported small cars were gradually increasing, information on GM's market share referred only to domestic production figures, so people could not notice the erosion of the domestic market by imported cars. This seemed incredible to us in Japan. I don't know if the cause was willful manipulation of information, obstinacy, or unawareness on the part of top management, but there was certainly a tendency to minimize the effect of imported small cars in Detroit and within GM.

In 1988, GM established a new imported-car franchise named GEO within the Chevrolet division. Until then, GM had sold the Nova (its version of the Corolla), manufactured by NUMMI, and the captive import Spectrum (its version of the Gemini), made by Isuzu in Japan, as well as the Suzuki Sprint (GM version of Cultus in Japan) through Chevrolet dealers. However, the impression for users was shallow, and dealers were puzzled as to how to sell them. The growth of sales was much lower than expected. In order to prop up sales, GM gave the unified brand name of GEO to all Japanese and Korean small cars: the Nova was renamed the GEO Prizm, the Spectrum the GEO Storm, and the Sprint the GEO Metro. All of them were sold through GEO. Thus imports were clearly differentiated from domestic Chevrolets.

The results were very successful, and sales reached 265,992 units in 1991 (*Financial Times*, January 18, 1992). If GM had created this kind of import-exclusive channel 40 years ago, I believe the state of the auto industry would be completely different today.

2) Innovation vs. mass production

Professor William J. Abernathy wrote in his book, *The Productivity Dilemma: Roadblock to Innovation in the Automobile Industry* (Johns Hopkins University Press, 1978):

> *Technological change, as Joseph Schumpeter has said, has proved to be a "gale of creative destruction," wrecking industrial lethargy and leading to improvements in productivity for the benefit of society. But in many cases, lo, as productivity increased, significant technological change became more difficult to achieve... As the product and the manufacturing process develop over time, costs decrease, product design becomes more standardized, and change becomes less fluid. At the same time, production processes, designed for increasing efficiency, offer higher levels of productivity, but they also become mechanistic, rigid, less reliant on skilled workers, and more dependent on elaborate and specialized equipment.*

Compared to the trends taking place in small cars all over the world, GM was consistently slow in introducing high-speed engines, five-speed manual transmissions, four-speed automatic transmissions, disk brakes, four-valve clean and fuel-efficient engines, and other components with new functions. As soon as rival Japanese automakers introduced new cars in Japan, GM and Ford purchased those cars in Japan and had them flown to Detroit, taken apart, and studied thoroughly. They even had a splendid dome where the disassembled parts and components could be displayed.

It is difficult to believe that GM engineers thought they could beat Japanese cars without introducing new technology and equipment. Even if they had, cost reduction, mass production, and higher productivity were given precedence, and technical innovation was pushed into the background. Until small cars began to be imported in large numbers, the oligopoly of the Big Three delayed the introduction of new technology that required a lot of capital investment.

In the early stages of the Saturn project, GM engineers and managers got together to come up with a plan: "Saturn is to go all the way — employ the most advanced technologies that can be found anywhere in the world. Saturn must be entirely new. There will be no carryover parts from other cars." It can be said that this slogan represented well the oppressed feeling among many at GM.

3) It is people who operate high-tech equipment

Those who have toured NUMMI in California and have seen TMM in Kentucky have said that there was nothing new technologically about their equipment and machinery. Compared to American plants, NUMMI and TMMK (Toyota Motor Manufacturing, Kentucky, Inc.) do not in particular have many robots or heavily automated equipment. Regardless, American workers operate the equipment, and both plants are ranked at the top among auto plants in the U.S. in terms of product quality and productivity. On the other hand, the Hamtramck plant, which was supposed to be the embodiment of high technology, could not realize the potential of the ultramodern equipment, and much of it was suspended or abandoned. Even today it has yet to go into full production.

GM established a high-tech factory as a means of lessening the gap in quality and productivity between them and the Japanese plants. They felt that by eliminating manual labor as much as possible and having machines perform every operation, quality and productivity would increase.

On the other hand, the philosophy at Toyota is that people are fundamental in increasing productivity and building quality. Robots and automation are utilized at Toyota to make it easier for the workers to do their job. Therefore, the priority is to first create a smooth production flow, then to think about how to automate it. As a result, each individual's capability is respected, and investment in the individual, in terms of training, is never spared.

A short article entitled "When GM's Robots Ran Amok" in *The Economist* (August 10, 1991) noted the following:

> *As one Hamtramck manager says: "There used to be a them-and-us mentality. Now we say: here's a part of business, help us run it." It was an expensive lesson, but GM eventually learned that it was not robots, but is own work force, that was its biggest and most valuable asset.*

4) GM's policy of supremacy at any cost, and corporate ills

It is natural that the employees of the world's largest automobile manufacturer should have a certain amount of pride. But if that pride goes too far — if they become convinced that what they are doing is always the best and that there is no need to imitate other people or listen to other opinions — it becomes arrogance.

One example of this arrogance can be seen in the problem of introducing quality controls. When Ford began a QC movement under the direction of professor Demming around 1980, as might be expected, GM also promoted a QWL (Quality of Work Life) movement throughout the company. GM believed that in order to produce high-quality products it was necessary to improve the environment of the workers. They purposely avoided expressing their movement directly as TQC or QC.

The QWL movement did absolutely nothing to improve product quality, as is obvious from the repeated failures I have described. However, there were a few other circumstances. During his tenure at Ford, professor Demming condemned GM's president McDonald at a conference. After that, the two of them were completely estranged, so that you had to whisper Demming's name at GM. It is not clear exactly what professor Demming said at the conference, but one might guess he suggested that the top must take the lead in the QC movement, and that McDonald could be criticized in that regard.

GM was stricken with typical corporate ills. When something undesirable happened, a committee would be set up to investigate it. Until the results were ready, vast amounts of labor and time were expended. Action would be taken on the results, but valid improvement was infrequent.

When John F. "Jack" Smith Jr. took the post of president, there were 11,000 white-collar workers at GM's head office. Through early retirement, redistribution and reorganization, this number was reduced to 2,000. For example, there were 1,600 people assigned to parts and equipment purchasing, but he reduced this to about 100. Through this kind of shrinkage, it is said that what was once the six steps of corporate structure has now shrunk to three. The fight against these corporate ills was president Smith's first job.

As mentioned previously, it is difficult to believe that when the plan for a major project such as the Saturn — costing some $50 billion — was announced, at least one executive vice president was not aware of it. It is also well known that chairman Smith's sudden organizational reform of GM — from Cadillac to Chevrolet, all the divisions integrated into two — created much confusion. The August 8, 1991 issue of *The Economist* reported that Smith himself admitted right before he retired that he should have discussed it with everyone.

GM's corporate ills are attributable to the shortcomings not only of the bureaucracy and middle management, as outside criticism would indicate, but also of top management, who must bear the brunt of the responsibility.

Chapter VI

The Rebirth of GM

The Rebirth of GM

VI-1. Coup d'Etat by Outside Directors

On April 6, 1992, the GM board of directors demoted president Lloyd E. Royce (then 55) to the non-director position of executive vice president. He was replaced by vice president John F. "Jack" Smith, Jr. (54), architect of the successful reconstruction of GM Europe, who was also given the post of COO (chief operating officer). Chairman Robert C. Stempel (58) stayed on but was replaced as head of the Executive Committee by an outside director, John G. Smale (64). Chief financial officer (CFO) Robert T. O'Connell (53) became chairman of GMAC, the sales and financing arm, and replaced as CFO by William E. Hoglund (57), previously in charge of the parts group. The abrupt announcement was described in the media as a coup d'etat by outside directors.

The events leading up to the unexpected changes can be summarized as follows:

1) In August 1990, president Stempel was appointed CEO (chief executive officer), replacing chairman Smith. He recommended his long-standing friend Royce as the new president. The board of directors was against the idea. They eventually agreed in the face of strong lobbying by Stempel, but stopped short of making Royce COO because of concerns about his capability for the post.
2) As the incoming chairman, Stempel was faced with the task of rebuilding the deficit-ridden North American Division. Unproductive plants had to be abandoned and the number of employees had to be drastically decreased in order to reduce fixed costs and bring down the break-even point. Ford and Chrysler had already cut fixed costs significantly during the downturn after the 1970s oil crises. By autumn 1991, however, Stempel had still not made any real move to get the rebuilding process under way. The North American Division had dragged GM overall into debt during the third quarter of 1990, and kept it there in spite of profitable performances by the overseas division and by such subsidiaries as

EDS (Electronic Data System), Hughes (Space and Aeronautical Machinery), and GMAC. On December 9, 1991, the GM board of directors convinced a reluctant Stempel to authorize a drastic rationalization program. The downsizing plan, involving the closure of 21 plants and the layoff of some 72,000 employees, was hastily announced. A list of 12 plants (including two assembly plants) was released on February 24, 1992, but even six months after the initial announcement the names of the remaining plants had still not been made public.

3) A *Fortune* magazine article entitled "Can GM Remodel Itself?" (January 13, 1992) said the following about GM's poor productivity levels: "By some key standards — how many worker-hours it takes to assemble a car, for instance — GM is an astounding 40% less productive than Ford. In 1991 GM lost, on average, $1,500 on every one of the more than 3.5 million cars and trucks it made in North America." Stempel announced a new plan to cut 9,000 white-collar employees during 1992, in addition to the earlier announcement of plant closures. Meanwhile, he promised that several new models would be introduced to the market (16 models were brought out during the 1992 model year). He also vowed to balance the company books during 1992 as the domestic economy began to recover, through a process of "evolution rather than revolution."

4) The continuing deficit was causing cash-flow problems. By the end of 1991, funds at hand had shrunk to just $3 billion, and it was feared that if nothing were done the balance would disappear altogether. Talk of bankruptcy began to circulate. Despite the downsizing announcement, financial sources downgraded the company's credit rating for long-term debt and rating on commercial bills. GM was obliged to take more drastic action.

5) Wall Street sources and pension fund managers were already saying that Stempel's moderate revolution was not enough, that the company books would not be balanced by the end of 1992, and that the deficit in the North American Division would have to be cut back to the bare minimum and a surplus brought in as quickly as possible. GM's outside directors, led by former Procter & Gamble chairman John G. Smale, covertly planned the boardroom coup d'etat.

6) Ira M. Millstein, counselor to the GM board of directors, played an active role behind the scenes. Millstein, who had been involved with GM since 1985, believed that outside directors had a duty to exercise their influence in the management of the company. Playing an active role in the coup, he made efforts to convince the outside directors that it was their duty to act immediately in view of the company's circumstances.

The new regime had begun, but the pace of company reconstruction under Stempel was still too slow. It seemed likely that the North American Division would post a massive deficit again in 1992. The pledge in April to achieve break-even within the year appeared impossible to fulfill even by the end of 1993. The value of ordinary shares released to finance a capital increase fell from $39 in April to less than $30 by mid-October. Doubts were raised about GM's ability to carry through its 1992 plan to close 21 plants and lay off 72,000 employees. Stempel was unable (or unwilling) to see the entire picture of the company crisis and too soft toward his employees because of his paternalism. Outside directors began calling for his removal.

Stempel's eventual resignation as chairman was accepted at a GM board meeting on November 2, 1992. Stempel had heard of the move to oust him and decided to step down himself. He was replaced by Smale, and president Jack Smith was promoted to the position of CEO. The separation of the offices of chairman and CEO was highly unusual. The president (Smith) now reported to the chairman (Smale), and thus the outside directors attained direct authority over the management of the company.

To ensure that the outside directors retained more permanent roles, and to prepare for the likelihood that an internal director would ascend to the post of chairman in the future, in March 1994 GM adopted "Twenty-eight Guidelines in Significant Corporate Governance Issue," basically a charter for the board of directors. The move was reported by Robert L. Simpson in the *Wall Street Journal* on March 28, 1994:

> *The guidelines embrace the concept of a "lead" outside director, call for thrice-yearly executive sessions of outside directors, and declare that board members shall "have complete access to GM's management," rather than being limited to funneling queries through the chief executive officer...*

> *The GM guidelines, first reported by the* Detroit News, *make it a board policy to designate an outside director to chair the three-times-a-year meetings of independent directors. John G. Smale, GM's non-executive chairman, currently fills that role. If a future chairman should again also be chief executive officer, then the outside directors would elect a "lead director"... The guidelines specify that the board make a formal evaluation of the chief executive each year. The non-executive chairman or lead director is to handle the performance review, which is to be based on "objective criteria including performance of the business, accomplishment of long-term strategic objectives, development of management, etc."*
>
> GM itself has yet to formally announce the guidelines.

Note: The term "objective criteria" is used in the Japan-U.S. Economic Framework Talks. But the interpretation of the term is different between the two countries. It is one of the reasons why it took so many days before they reached an agreement. GM shareholders and the outside directors who represent them have no qualms about using the term to evaluate the performance of serving directors. However, in the context of the talks, it is hard to believe that a foreign government, namely the U.S. government, is going to evaluate the behavior of another government on equal terms using objective criteria.

Only time can tell if the new regime is successful or not. But I believe that such a large enterprise as GM should have a mechanism for assessing the performance of the CEO, who has authority over all aspects of company operations. (Auditors are mainly concerned with auditing the results of corporate activity; they are not properly equipped to monitor corporate activity itself.) In today's society, it does not make good business sense to give to a single person, such as a chairman or CEO, full responsibility for operating the company, making decisions that affect the fate of the company, the shareholders, tens of thousands of employees, and hundreds of dealers as well as suppliers.

In Japan, some manufacturers should have changed their policies from irresponsible over-investment to more tightly controlled practices before the bubble economy burst. Instead, some continued their massive investments by obtaining funds from overseas equity markets at low interest rates or effectively zero interest. Some general manufacturers are rumored to be having difficulty recovering. This might be attributed partly to the absence of a mechanism for monitoring the performance of top management, since main banks lost their initiative to their debtors due to the easy money policy. It might be the time for outside directors to be introduced to monitor CEO performance in place of the financial institutions.

VI-2. Management Reform Under President Jack Smith

The most pressing task for incoming president Jack Smith was to rebuild the debt-plagued NAO (North American Automobile Operations). Upon his appointment, Smith announced that he was taking personal control over NAO to balance the accounts during 1993 (excepting depreciation, interest payments, and taxes). He then pushed through the following personnel changes:

1) After taking office, Smith immediately assumed chairmanship of the NAO Strategy Board. It was set up in December 1991 under former chairman Stempel to oversee the reconstruction of NAO. Smith completely reshuffled the board, appointing younger, talented staff regardless of seniority. He moved his office from Detroit to a corner of the Warren Technical Research Center 30 miles northeast of Detroit, arrived at seven o'clock every morning, and conducted Strategy Board meetings there. Before then, GM's top executives rarely set foot outside their offices on the 14th floor of the head office in Detroit, from where they ran the company.
2) J. Ignacio Lopez de Arriotua, vice president of GM Europe and purchasing manager for Europe, was appointed worldwide purchasing manager and given the additional post of vice president at the head office in the U.S. He brought to NAO the tough cost-cutting strategies he had applied successfully to the purchasing division in Europe.
3) With the resignation of chairman Stempel and the appointment of president Smith to CEO in November 1992, GM Brazil president G. Richard Wagoner Jr. was promoted to CFO (chief financial officer) at the age of 39. Replacing the 58-year-old Hoglund with an officer 20 years younger injected a breath of fresh air to the company. The new president of GM Brazil, Mark T. Hogan, was a second-generation NUMMI controller (accounting manager) and well versed in Toyota production methods.

4) President of GM Europe Louis R. Hughes was given the additional post of executive vice president in charge of overseas operations at the head office, to permit Smith to devote his full attention to NAO.
5) The technical organization of the company was completely overhauled and condensed into the North American Passenger Car Platform and Truck Platform departments. The Vehicle Launch Center was set up within the North American Passenger Car Platform department to inspect all new vehicles.

While pushing ahead with economic reform under the new regime, GM also strove to improve hostile relations with the United Auto Workers (UAW). The company arranged a meeting with the union to explain its predicament. Although GM didn't succeed in holding down wages in the parts divisions when the agreement on working conditions was renewed, it was allowed to lay off workers on the condition that they could work at other factories within a 50-mile radius.

The UAW also approved the sale of parts factories, although only in the context of continuing massive staff cuts. The union has not softened its stance as such; two strikes at a number of plants during the third quarter of 1994, for instance, scuttled a planned production increase. In general, though, the approach taken by Smith thus far suggests that the workers issue will not pose a major obstacle to reconstruction.

VI-3. Overhaul of Product Planning

Use of New Technology

As mentioned in the previous chapter, GM had considerable technological and development expertise but did not apply new technology quickly enough to vehicle design. One example is the introduction of multi-valve engines.

The traditional four-stroke internal combustion engine had always had separate intake and exhaust valves on each cylinder. During the late 1970s, tests showed that multi-valve (three or four valves on each cylinder) engines produced better fuel economy and cleaner combustion. Yet the first GM-made four-cylinder 16-valve engine did not appear until August 1987. Toyota brought out a six-cylinder 24-valve engine in August 1982 and a four-cylinder 16-valve engine in May 1983. These were followed by an eight-cylinder 32-valve engine in 1990. By 1992, all leading Toyota engines were modified to multi-valve configuration. (Only the Toyota Century, a large-scale passenger car for the Japanese market with annual production under 1,000 units, remained a twin-valve).

GM announced that the 1991 Cadillac would be equipped with the new eight-cylinder, 32-valve North Star engine. But the performance of the new engine was said to be unsatisfactory, and their production rate was not steady. In October 1994, GM disclosed that computers controlling the North Star engine and automatic transmission would henceforth be obtained from Siemens in Germany instead of Delco Electronics, a subsidiary of GM Hughes Electronics under the umbrella of ACG (Automotive Components and Parts Group of GM). The switch to an outside subcontractor for the computer controller, a pivotal part of the car that governs not only the engine but the way the car feels in general, aroused widespread interest both inside and outside the company.

These delays prevented GM from achieving the CAFE (Corporate Average Fuel Economy) standards with 1994 models. In an announcement in May 1994, GM reported that it had been unable to meet the standards for either passenger vehicles (27.4 mpg, just under the required 27.5 mpg) or light trucks (19.9 mpg versus the required 20.4 mpg). Ford and Chrysler vehicles comply with the standards.

Although slow to tackle improvements to the basic engine structure, GM was very enthusiastic about introducing new and improved parts that did not require significant changes to equipment and facilities. GM led the field in cruise control, for instance. It was the first to install dial-type cruise control (in 1961), followed by a one-touch-resume electronic model (1965), analog control (1968), one-touch analog digital electronic control (1974), the ECU actuator (1989), and an ECU fully integrated with the engine controller (1990). When it comes to product improvement that can be achieved without major investment in plant and equipment, GM has kept abreast of all the latest technological developments.

More recently, GM has been quick to fit a large number of models with ABS (antilock braking system), a prime example of an electronic controller.

The real issue is model changeover cycle. In the past, Japanese automakers were unable to keep up with their American counterparts, who released new models every year. Instead, Japanese manufacturers would make a complete changeover of car line every four or five years with an additional, minor change every two years or so. It was not so long ago that the industry and journalists in Japan realized that the actual model changeover period was nearly two times longer in the U.S.

At the time, we believed that a full model change was required every four to five years in order to bring customers the latest in automobile technology. In the U.S., automakers will continue the production and sales of new models until their market value and technological value decrease to virtually nothing. Their production facilities and equipment are kept in operation as long as possible. When the end finally comes, not only the car models but in some cases the factories, equipment, and even workers are abandoned. It's like slash-and-burn agriculture. The Japanese were more than surprised by this behavior in the U.S. auto industry. Meanwhile, Japanese automakers have been accused of keeping their model change cycles too short as part of a so-called planned obsolescence policy. Of course, this is not true.

Note: The model changeover cycle is not directly related to the life of the vehicle. After being exchanged for a new model, a secondhand vehicle passes through many hands before reaching the end of its life span (which in Japan depends largely on the cost of registration and insurance of an older vehicle versus the cost of purchasing a newer one). An appropriate model changeover cycle provides the market with a steady flow of secondhand vehicles and enhances the overall quality of vehicles on the road, contributing to reduced fuel consumption and low air pollution. In the U.S., large, polluting, secondhand gas-guzzlers still account for some 10% of all cars on the roads.

Surely it is the duty of vehicle manufacturers to embrace new technology to improve vehicle safety; reduce environmental pollution; boost fuel economy; enhance reliability, durability, and handling; improve performance and driving comfort; and make cars cheaper and easier to obtain. It is no exaggeration to say that the humiliation of the Big Three has been brought about by their failure to fulfill this duty. Now, of course, they are struggling to shorten development periods for new models and ensure problem-free production startup.

Products for the Lifestyle Market

Professor Peter F. Drucker wrote an article in the *Wall Street Journal* (June 18, 1991) titled "The Big Three Miss Japan's Lesson." He said that the marketing theory developed by GM's Alfred Sloan — according to which products should be developed for four or five separate socioeconomic groups — had failed and had changed to the current era of the "lifestyle market." The Big Three, however, had overlooked this trend. He said that the current situation has changed, and products have improved and are being offered at the lowest price possible. Yet the domestic share is being eaten up by the Japanese. Drucker went on to caution that this lifestyle market is not stable, and because it changes continually, American automakers must develop and supply products quickly like the Japanese if they are to take advantage of this trend.

In 1981, GM began development of its front-wheel-drive GM10 series as part of the program to change its core segment of intermediate cars, A- and G-bodies. Its goal was to produce 2 million units of the GM10 series.

As a result of GM's reorganization, this program was repeatedly delayed. Meanwhile, A-body sales were doing relatively well; therefore, GM decided to leave it a rear-wheel-drive car. The G-body, on the other hand, underwent a complete model change and became a front-wheel-drive GM10. The new GM10 consisted of the two-door coupe, four-door sedan, and station wagon models for Chevrolet, Pontiac, Oldsmobile, and Buick.

In June 1986, GM announced that its sales goal for the GM10 would be 1.7 million. However, it soon reduced its sales target to 1 million units, as a result of many factors. With the successful introduction of the Ford Taurus and Mercury Sable, GM had quickly changed its styling, and with the success of the Chrysler mini-van, it had decided to abandon the station wagon. Further, GM greatly reduced its initially announced investment figure of $7 billion to save costs. With the reduction of the sales target, the two-door coupe was finally introduced in 1987, and a year later, the four-door was introduced. In an article in the *Nikkei Sangyo Shimbun* (October 30, 1989), Maryann Keller estimated investment costs at more than $6 billion. This project took nearly nine years.

The result of the GM10 program can be seen in Table 11. The results were far from successful.

Table 11: Sales of GM Cars by Platform Before and After GM10
(No. of Units Sold)

	1991	1985
A-body		
Chevrolet Celebrity	616	363,619
Buick Century	128,904	234,508
Oldsmobile Cutlass Ciera	135,670	333,585
Pontiac 6000	21,624	165,728
Total	286,814	1,097,440
G-body = GM10		
Chevrolet Lumina	217,555	112,585
Pontiac Grand Prix	103,203	57,153
Oldsmobile Supreme	87,540	217,504
Buick Regal	104,802	112,590
Total	513,100	499,832
Grand Total	**799,914**	**1,597,272**

Source: *Automotive News* and *Ward's Automotive Reports*
*Figure is for Chevrolet Monte Carlo

By 1991, when the recession had bottomed out, total sales of GM vehicles had fallen to 63.1% of 1985 levels. Total sales of the A-body and B-body vehicles, which were key vehicles in the GM10 program, were reduced by half, having largely exceeded the average. Furthermore, everything that Drucker had pointed to as flaws in the American auto industry was proved in this project:

(1) Because it spent nine years on development, GM was unable to keep up with the changes taking place in American lifestyles. The decision not to go ahead with the full model change of the A-body was probably the first mistake. The initial plan was scrapped because the current model was doing so well. GM failed to recognize several competitive trends: the successful introduction of the Ford Taurus and Mercury Sable, the high status of imports, and the failure of the old A-body model to fit customers' lifestyles. More importantly, GM failed to recognize that the diversity in lifestyles severely decreased the market potential for a particular model.

(2) GM blunted the initial impact of the GM10 by releasing the two-door coupe ahead of the four-door version at a time when four-door vehicles represented the major area of demand.

(3) Sales figures for 1991 show that not one of the plants producing the GM10 maintained a manufacturing volume sufficient to turn a profit. The figures are higher for Chevrolet Lumina, but it is built in two plants, No. 1 and No. 2 plants in Oshawa, Ontario (Buick Regal is also built in the latter plant). The Pontiac Grand Prix is built in Fairfax, Kansas, and the Oldsmobile Supreme is built in Doraville, Georgia. So the total number of assembly plants producing the GM10 is four, and all are supplying in total 500,000 cars. Here one can see the severity of the problem. Since the Framingham plant, the model of inefficiency, closed, all A-body cars are now produced in Oklahoma City. Production of the Celebrity has come to an end, and production of the Pontiac 6000 was scheduled to end in June 1992. It is interesting to note how large an effect the failure of one model had on the entire program. At the same time, one can fully understand how the introduction of a production system capable of mixed production of several models at a plant became essential and indispensable for succeeding in a more fragmented market.

The failure of the project shook the very foundations of the company. Despite the reforming efforts of president Smith and those around him, GM is still having considerable difficulty in the aftermath of its failure to keep up with the lifestyle market.

The successor to the GM10, GM's next flagship passenger car, will not be released until 1997. During this period, the earlier Chevrolet Monte Carlo 1995 two-door coupe model will be revived, while the Chevrolet Lumina four-door sedan will be given a facelift, including some new exterior panels. But most observers believe these changes will have only a limited effect against strong competitors such as the Ford Taurus, Honda Accord, and Toyota Camry.

The last few years have seen spectacular growth in the mini-van market created by Chrysler. Following concerted efforts to make up for lost ground, Ford is now enjoying the benefits of the mini-van trend together with Chrysler. GM, meanwhile, has failed in this field also.

The Chevrolet Lumina APV was released at the end of 1989, just as the mini-van boom was taking off. It was poorly received, not least because the front end stuck out so far that the driver could not see the front of the vehicle. There were also complaints about the shape and position of the tail-lights at the rear. Sales were so bad that at one stage GM considered halting production of 1995 models at the Tarrytown plant in New York and closing down the plant altogether. Despite a pronouncement from Smith that product development should focus as a matter of urgency on a new mini-van model, there was no way it could be brought onto the market by the original 1995 deadline. It would have to be the end of 1996 at the earliest. Front-end modifications to the existing APV in 1992 had met with a degree of success, and sales had improved slightly in line with the general economic recovery and the boom in the mini-van market. The Tarrytown closure was put back ten months.

In addition, GM did not have a jeep-type recreational vehicle (RV) like the Ford Explorer and Chrysler Cherokee. Despite deficiencies in the passenger car and RV product lineup, GM still holds top position in terms of production and sales of bargain-priced "low-tech vehicles" such as large pickup trucks equipped with massive engines, where it has always done well.

(The term "low-tech car" was coined by Professor David E. Cole of the University of Michigan during an address to the International Economic Institute in 1993 to describe those models that had been left behind in the race to employ advanced technology.)

GM's share of the market is slowly but surely starting to decline; in 1994 it only barely managed to maintain the previous year's level of 33%. Some observers say that GM's recovery of its market share will depend on whether the present booming economy can be maintained through 1997, when the new product lineup will be completed.

Meanwhile, from the 1980s through the early 1990s, Toyota was unable to produce a highly successful product in the domestic market, resulting in a slight drop in overall market share. Like GM in the U.S., Toyota was slow to recognize the emergence of the lifestyle market in Japan (mimicking the U.S., only two or three years later), and consequently lost the lead to rivals such as Mitsubishi, Suzuki, and Fuji Heavy Industries. However, with the RAV4, a small, four-wheel-drive SUV released in June 1994, Toyota suddenly had a hit product on its hands.

Fewer Vehicle Platforms

Right after assuming the presidency, Smith had worked hard at the NAO Strategy Board to develop a plan to reduce the number of platforms in conjunction with the review of product planning. The plan aimed to revamp costly, unproductive structures by eliminating surplus staff and production equipment, thereby shortening the development period and reducing the overall cost of developing new models. According to media reports (*Ward's Automotive Report*, November 16, 1992), Smith announced during a November 1992 meeting with analysts in New York that the number of passenger car platforms (then 12) would fall to 11 by the end of the year, then to 10 by 1995, nine by 1996, and five by the year 2000. The five would be subcompact, compact, intermediate, regular, and the Y-platform for the FR vehicle.

Only by strong leadership could incoming president Smith institute this unprecedented program. His predecessor, Lloyd Royce, had not considered using fewer platforms or streamlining the product development process. General managers of the five sales channels, meanwhile, were more concerned with preserving the life span of their own models and with insisting on a variety of products. GM is currently working on plans to introduce the Opel from GM Europe into the NAO to enhance new model development and simultaneously improve development efficiency throughout the company.

A plan for reducing the number of platforms for light (small and mid-size) trucks is also persuasive. It is assumed that pickup trucks are developed in parallel with SUV's (sports utility vehicles), integrating them into three platforms such as large, medium, and small. A small pickup and a medium SUV (both from the small platform) will be released by making "skin changes" (minor modifications to external panels only) as the 1994 and 1995 models respectively. Although new models of large pickups and vans will not appear before 1997, GM will be able to have a product lineup by 1996 for the market where it has traditionally been weak. The plan for introducing a medium-sized SUV from Isuzu, of whose stock GM owns 37.5%, is being carried out, too.

VI-4. Production Startup Problems

GM suffered from not being able to start the production of its 1995 models smoothly. The initial production of the Oldsmobile Aurora, which was a symbol of the revived Oldsmobile line, and the Monte Carlo, which was introduced to refresh the Chevrolet line, was rather chaotic, with the result that GM could not capitalize on their new marketing idea.

Toyota experienced the same production startup problems in the 1960s. The Corona RT40, which overwhelmed the Datsun Bluebird, suffered from startup problems. Because of the poor quality of the body assembly, the first 500 units could not be sold on the open market but instead were sold to employees at a special marked-down price. I was one of the lucky employees; despite some blemishes on the external finish, the driving performance of my new RT40 was perfect in all respects. Even in today's terms it was an excellent bargain. In the ensuing 30 years, Toyota has never made the same mistake again. It is thus somewhat irritating to witness the production startup problems in GM as recently as in 1994.

In a *Business Week* article (July 11, 1994), James B. Treece reported:

U.S. carmakers still lag far behind their Japanese rivals in changing over their factories to produce a new model. Production preparations for the 1995 Chevrolet Lumina began on November 19, 1993, but production didn't start before February 14, 1994. Output was zero for 87 days. When it finally started up again, the factory made just 10 cars a day and painted all of them white. By April 1, it had built a total of 288 units. The plant won't reach its full-line speed of 60 cars an hour until August, though its palette has expanded to 7 of the 10 colors it will produce. Production of the new Oldsmobile Aurora and Buick Riviera started slowly to maintain quality. Daniel Blaskovich, general manager of Tom Blaskovich Chevrolet-Oldsmobile Inc. in East Chicago, Indiana, told me, "I'm afraid we'll lose four months' sale with it, too."

The article went on to list the length of production shutdowns for model changeovers at other manufacturers: three days at the Marysville plant in Ohio for the 1994 model Honda Accord, 18 days at the Georgetown plant in Kentucky for the 1992 model Toyota Camry, and 60 days at the Kansas City plant for the 1995 model Ford Contour and Mercury Mystique.

Production startup delays cost GM 35,000 units during the second quarter of 1994 (*AP-DJ*, 28 July 1994), not to mention 90,000 units during the third quarter of 1993 during the startup period for 1994 models.

The problems with startup production of new models still had not been solved by early 1994. This, coupled with two strikes, caused GM's business performance for the third quarter of 1994 to fall below expectations. On the day the figures were announced in October 1994, GM shares plummeted by $3.75. Although GM is endeavoring to improve production startup, analysts remain skeptical about the outcome due to the age of plant equipment and facilities and the lack of technical expertise.

The *Wall Street Journal* (October 21, 1994) predicted that the startup problem would persist through the last quarter of 1994 and into 1995.

Toyota, meanwhile, had been working to develop new production methods as an extrapolation of its current approach. In a report entitled "Development of a Flexible Production System for Body Welding" (*Journal of SAE of Japan*, January 1991), four Toyota engineers from the No. 2 Production Engineering Division (Hideaki Tobita, Masuji Arai, Youichi Shibata, and Hirobumi Muda) wrote about Toyota's advanced FBL (flexible body line) automatic body assembling system. Under the new FBL method, several different car bodies could be manufactured on the same line, enabling model changes without disruption to the production process. The report also claimed that:

> *...already, one model changeover and three model additions have been made without a single day's stoppage. Availability directly before and after the changes has not been significantly affected, and in all cases production had returned to 95% of normal levels within two to three weeks. FBL also cut in half the preparation man-hours, lead time, and investment for plant and equipment normally required for a model change.*

GM's production startup problems were caused by the inflexibility of the production equipment setup as well as limited floor space for the preparation for building new models while the production line of old models was in operation. But the biggest problem was the shortage of production control experts.

In November 1993, Opel's Eisenach plant president, Tom LaSorda, was recalled to the U.S. and appointed manufacturing manager for the Cadillac/Luxury Car Engineering and Manufacturing Division. LaSorda was the vice president of CAMI, a joint venture between GM and Suzuki in Canada. In 1991, he had been chosen to supervise the overhaul of the Wartburg plant at Eisenach (which mass-produced passenger cars for the former East German market) and preparations for building the new Opel Astra. His efforts to install Toyota production methods he had mastered in Canada (including frequent visits to TMMC) evidently paid off: The production startup was extremely smooth. By 1993, the Wartburg plant was acclaimed the most efficient in Europe. Representatives from automobile plants throughout the Continent flocked to marvel at LaSorda's spectacular turnaround. Yet his success only served to demonstrate the dearth of superior production control managers elsewhere at GM.

At the initial stage of production, the main concern must be quality rather than quantity. The priority given to quality assurance over volume during production startup made the Saturn famous. The same approach appears to have permeated through to other models. GM vehicles have started to rank highly in the J. D. Power survey of initial quality levels. In spite of problems with production startup, GM is obviously trying to give top priority to quality.

Chrysler, meanwhile, has been forced to recall a number of models, such as the new Neon. The J. D. Power quality survey of May 1994 ranked all Chrysler brands (Plymouth, Dodge, Chrysler, Eagle, and Jeep) as average or below. Chrysler chairman Robert Eaton called off the June test-drive session for executives, a regular event for new models, and sent out a stern demand for quality improvements. Production of the new Chrysler Cirrus and Dodge Stratus, touted as direct challengers to Japanese imports, was put off until quality issues could be resolved.

VI-5. Lean Production

In the January 13, 1993 issue of *Fortune*, Alex Taylor III wrote about the inferiority of GM's productivity compared to Ford's.

> *According to an internal GM analysis, 20 of its car assembly plants require an average of 35.7 worker-hours per vehicle, vs. 21.6 hours for Ford. Worse, GM needs 32.2 hours to 36.3 hours to put together such midsize models as the Chevrolet Lumina and Pontiac Grand Prix, while Ford can turn out the similarly sized Taurus and Mercury Sable in under 17.2 hours. Figuring labor at $31.50 an hour for wages and fringes, overall Ford has a $441 cost advantage on the factory floor alone.*

This staggering difference in productivity can be attributed to three factors. First, since Ford had conducted a massive reduction of the staff and factories during the recession triggered by the 1970s oil crises, it was much leaner than GM by the beginning of the 1980s. Second, Ford had been taking a keen interest in the efficiency and productivity of the Japanese automobile industry since the mid-1970s, sending representatives to Japan, studying Japanese production techniques from sources such as academic research findings, and using those methods deemed suitable to its needs — for instance, the team system. And third, Ford chairman and CEO Donald E. Petersen himself had taken the initiative to promote Dr. Demming's Statistical Quality Control Management throughout the company.

GM has been making efforts to carry out the same sweeping reduction of factories and staff implemented over a decade earlier by Ford. Concerning the second and third actions by Ford mentioned above, GM has not been completely inactive. The problem has been its failure to implement changes on either a systematic or a company wide basis as Ford had done. Consequently, product quality improvements are now the responsibility of each manager, and quality and productivity vary according to each plant. For instance, the Oklahoma City plant is said to be one of the best GM plants. Gregory A. Patterson wrote in the *Wall Street Journal* on August 29, 1991:

> *What's different is that Oklahoma City's employees are radically changing how they work. They're putting an American face on a manufacturing system developed by Toyota and other Japanese rivals. This system, known inside GM as "synchronous manufacturing" and also called "lean production" or the pull system, emphasizes doing more with less.*

Jack Evans, the Oklahoma City plant manager, got a look at Toyota's highly disciplined systems in 1987, when he and some Oklahoma City employees toured NUMMI, which Toyota manages. Now it's Mr. Evans who gets visitors from other GM factories.

Patterson went on to say that although GM has such an outstanding plant, on the other hand it also has a plant in Orion Township, Michigan, which, although it has more than four times the robots of the Oklahoma City plant and is five years younger, is ranked the lowest in quality among GM plants.

In February 1992, I had a chance to meet with R. D. Anderson, manager of the Ford Atlanta plant, in Tokyo. He smiled and said, "We have a secret that there is no secret [for our success]." He said that all they are doing is making an effort to respect the rights of the workers and have good relations with them. The differences among GM factories are related to the quality of relations between management and labor. Management needs to understand the circumstances of labor, and vice versa.

When it established the Saturn plant as an independent subsidiary, GM had tried to build the most efficient facility possible based on an idealistic labor-management relationship. Its strategy was to push for improvement in conventional plants by pointing to Saturn as a model plant. Although one can't claim that it completely ignored the capability of the ordinary union worker at the existing plants, one can say that it failed to improve its labor-management relationship because it never fully respected the capability of its workers.

As a GM watcher who is interested in whether this experiment could rebuild GM, I was concerned about the wisdom of building a completely new factory, employing new workers, and treating the operation as a separate company. In the end, G. Richard Wagoner Jr., the president of the NAO, decided in September 1994 (three months after his job assignment) to merge the Saturn, supposedly independent from GM, into the Small Vehicles Section of the NAO.

The book *Made in America* (Michael L. Dertouzos, Richard K. Lester, Robert M. Solow, and the MIT Commission on Industrial Productivity, MIT Press, 1989) stated:

> *In firms organized according to the logic of mass production, workers are treated as replaceable parts. By defining jobs narrowly and making each job relatively easy to learn, American industry pursued flexibility through the interchangeability of workers with limited skills and experience rather than the cultivation of multi-skilled workers. Employees could be hired and fired with the ups and downs of the business cycle without much loss of efficiency. The result was a progressive narrowing of worker responsibility and input and the tendency for management to treat workers as a cost to be controlled, not as an asset to be developed.*

If this type of old thinking is still in place, then the road to becoming a lean producer is a very long one.

Lean production will eventually come to GM's operations in North America; the question is how long that will take. GM chairman Smith has stated his intention to convert all GM plants to NUMMI production methods by 1996.

VI-6. Enhancing GM's Technological Potential

The GM Technical Center in Warren, on the outskirts of Detroit, is a complex of seven buildings on 3.6 million square meters, built from 1949 to 1956. In 1969, the center had nearly 5,000 scientists, engineers, and designers. According to *History of the American Automotive Industry* by Shoichi Inoue (Sanyo Tosho Shuppan, Japan), their efforts were concentrated mainly in the following areas:

1) Energy conversion (rotary, starling and gas turbine engines, fuel cells, electrochemical cells, and lightweight, high-efficiency electric motors)
2) Improving pollution control and safety (reducing emissions of harmful gases from the internal combustion engine and developing a practical electric vehicle)
3) Reducing lead time using numerically controlled machine tools

Unfortunately, precious little of the new technology developed at the center has been put to use.

Convinced that the rotary engine (RE) revolution was inevitable, then-president Edward E. Cole bought patent rights for the RE in 1970 from three companies, Audi-NSU, Wankel, and Curtis Wright.

His aim was to put in place by 1974 a production structure capable of manufacturing one million RE vehicles annually, and eventually to equip all GM models with rotary engines. In response to concerns about the RE's compliance with stricter emission standards, Cole stated that GM had solved all fuel economy and emissions problems of the Wankel (rotary) engine.

In spring 1973, GM announced that an RE-equipped Chevrolet Vega was in the final stages of development and that parts were already being ordered. Then, at the end of the year, GM indefinitely postponed the RE for the 1975 model Vega, citing problems with fuel economy. In September 1974, just one week before retiring on the age limit system, Cole announced that the small sports car model scheduled for release in spring 1975 would no longer carry a rotary engine. To this day, GM has had no success in its efforts to develop the rotary engine.

Mazda, known worldwide as a leader in the development of the rotary engine, stayed ahead by suppressing peripheral patent rights. Other manufacturers who sought to catch up were unable to produce an RE vehicle from the NSU's patent alone. Toyota committed R&D efforts to develop its own RE engine, and Toyota engineers claimed that they had succeeded in making a better one than Mazda's. Toyota negotiated with Mazda over the use of peripheral patents but to no avail.

MITI tried to persuade Mazda to change its mind, citing the impetus given to the audio industry worldwide when Phillips, a Dutch company specializing in audio engineering, released its patents on the compact cassette. It is assumed that GM also asked Mazda for technical assistance, clearly with no more success.

While Toyota moved quickly to abandon the RE vehicle in view of the patent wrangling and continuing problems with fuel economy, GM stuck it out, suspending the project when production preparations were already under way. Some observers say that this meant the downfall of Cole, the man held ultimately responsible for the project.

With the gas turbine engine, GM planned to begin trial production of 60 units in 1971, followed by commercial production in autumn 1972. Once again the plans were abandoned at the last moment. Toyota's own gas turbine engine for small cars, exhibited in the Corolla and Crown at the 1983 Tokyo Motor Show, was never fully developed due to problems with exhaust gas and the strange sounds emanating from the turbine itself.

GM's success with research on electric vehicles spurred legislation in California requiring manufacturers to offer zero-emission vehicles by 1997. As the deadline approaches, however, the industry is calling for an extension or at least relaxation of the law, because of spiraling costs and engineering difficulties.

At the time, GM believed that all the projects proposed by their research center could be developed for practical use within the 20th century. It now appears unlikely that any of them will be. Toyota tackled similar projects and met with similar difficulties. Our experiences indicate how difficult it is to predict the outcome of research and development in this field.

Since the 1930s, GM had consistently led the industry in automotive technology, for instance with the development of the automatic transmission. After World War II, it continued to tackle new engineering fields with renewed resolve. The failure of the Corvair, described in Chapter V, was one of the company's more daring technological ventures. The Vega, although also a victim of poor quality, contributed front-wheel-drive technology that was subsequently used worldwide in FF vehicles.

Similarly, although the vehicle production facilities at the Hamtramck plant are not yet fully utilized, one has to applaud GM's bold initiative in trying to build an ideal factory. The desire to continually take on new challenges is evident also in the unique design of the Saturn and its production systems. This corporate spirit—remaining undaunted in the face of failure with a model or two or even a new production system—has enabled GM to handle such enormous risks when taking on technological challenges.

Under the early retirement incentive scheme for white-collar employees, introduced in 1974, a number of leading engineers have left to form engineering consultancies which then take on GM contracts. Some observers say that GM is no longer capable of developing vehicles without the help of these outside companies. As production technology becomes more and more advanced, engineers are required to work on the plant floor where production takes place. The importance of these engineers has been underestimated in the U.S.; even GM has never had a sufficient number of them. The best, most talented engineers have also been absorbed since the 1960s into the space engineering and other massive research and development fields. Automobile companies are now suffering from a shortage of skilled engineers.

In 1993, I asked professor David Cole of the University of Michigan about GM's alleged shortage of engineers. Cole, son of Edward E. Cole, GM president from 1967 to 1974, denied this and spoke of hundreds of top engineers working at various GM installations. Yet in spring 1994, GM publicly confirmed the shortage and announced plans to recruit 2,000 new engineers. A significant number of American engineers working for Toyota and other Japanese manufacturers in the U.S. promptly switched camps. (Fortunately, we received a flood of applications from skilled engineers and so were spared a potential recruiting headache.)

VI-7. Independence of Parts Suppliers

Since early 1991, as Japanese transplants in the U.S. steadily increase production, the rapidly worsening imbalance of the auto parts trade between Japan and the U.S. has prompted serious discussion in the U.S. A prime example is a report entitled *The U.S.-Japan Bilateral 1994 Automotive Trade Deficit*," issued in April 1991, and written by Sean P. McAlinden, David J. Andrea, Michael S. Flynn, and Brett C. Smith of the Office for the Study of Automotive Transportation, Transportation Research Institute, University of Michigan.

Some have insisted that Japanese manufacturers should purchase parts and components made in the U.S. not only for use in their transplants but also for assembling of cars in their factories in Japan. A report titled *The Future of the Automobile Industry — It Can Compete, Can It Survive?* (by Clyde V. Prestowitz, Jr. and Paul S. Willen, Economic Strategy Institute, 1992) was a typical example of this argument. Since NUMMI had begun operation, Toyota had made efforts to purchase U.S.-made auto parts. Chairman Shoichiro Toyoda of Toyota Motor Corporation personally directed the company to boost local purchasing in response to these demands. Toyota held many meetings throughout the U.S., calling on suppliers to sell parts to Toyota, and distributed leaflets explaining in great detail how to set up a business relationship.

In spite of these efforts, the number of new-parts purchasing contracts has not increased. One of the main reasons is that there is a big difference between U.S. and Japanese manufacturers in their parts purchasing policies. According to MIT's survey report, *The Machine that Changed the World*, in-house procurement is 70% at GM compared to just 27% at Toyota. (The founder of Ford, Henry Ford, was legendary for his pursuit of total in-house procurement, right down to the raw materials. He even built ironworks and bought cotton fields to produce textiles for internal fittings. His third successor at the helm, Henry Ford II, began to reduce in-house procurement in the 1950s, and the ratio was down to around 50% by the 1980s.)

In Japan, the government had made efforts to promote the automobile parts industry since prewar days. MITI adopted a policy assisting the establishment of the auto parts industry as one of its methods to revive the domestic automobile industry, then in straitened circumstances brought about by prolonged labor strikes in the early 1950s. The parts industry could get financial aid from the government fund. Meanwhile, automakers were advised by financial organizations such as commercial banks and various government finance corporations to become lean by separating sales functions as well as parts and components manufacturing. This, too, contributed to the development of the industry.

Auto parts manufacturers that became independent from automakers expanded their dealings to automakers other than their parent companies and thereby strengthened their financial foundation. Through fierce competition, auto parts manufacturers with excellent technical expertise and a strong financial base have grown. As mentioned above, this is one of the main reasons Japan still has 11 full-fledged automobile manufacturers, eight of them making passenger cars. The big companies as well as the medium-sized have survived with the support of these parts manufacturers.

The high rate of in-house parts procurement by U.S. automakers prevented the growth of the parts industry in this country (I suspect that in prewar days the big manufacturers attempted to squeeze out new entrants by holding their own parts production functions within them). As a result, in 1990, there were only about 30 auto parts companies in the U.S. capable of matching their leading Japanese counterparts in terms of technical expertise and accumulated capital. Japanese automakers thus cannot significantly increase procurement of parts and components made in U.S.until the in-house parts divisions of the Big Three start to supply parts to outsiders.

Since the in-house parts division is not open to outside competition, the transfer prices to other divisions within the company are calculated based on the manufacturing costs. As the division has no real incentive to bring down costs, it is generally unproductive. Therefore, the rationalization of parts manufacturing inside the company has become one of the main tasks of the management at GM.

Speaking with industry representatives in Washington, New York, Boston, and Detroit in autumn 1991, I argued that as part of the reviving process GM should divide its parts division into several independent affiliated companies. But most of them said it was impossible because GM had to get UAW approval when changing in-house procurement over to outsiders and also had to guarantee the income of surplus workers (UAW wages are twice those at ordinary parts factories). When I talked to the president of Detroit Diesel, Roger S. Penske, he wrote down on the notepad in front of me a string of specific proposals on how to separate GM's parts division. When I visited Dr. James P. Womack of MIT, he would repeat my intriguing suggestion to GM president Stempel, whom he was due to meet the following day. Maryann Keller, meanwhile, told me in New York, "The idea is totally out of the question. The union wouldn't allow it. GM already has plenty of other subsidiaries it would like to get rid of, and they can't even do that."

My meeting with professor David Cole of the University of Michigan in October 1991 turned into quite a debate. The Japan-U.S. automobile-parts trade issue was initially brought to the forefront by an analyst from his own Transportation Research Institute of the University of Michigan, while Cole himself is one of the leading voices protesting the trade imbalance in parts trade. Our discussion was therefore confined to a rather narrow range of issues. According to professor Cole, the U.S. parts industry can be divided into three tiers. About thirty first-tier suppliers are excellent companies that promote technical development internally and insure high productivity and quality. Manufacturers in the third tier, the overwhelming majority, operate more like "job shops." Those in the second tier, midway between the upper and lower levels, are doing their utmost to move upward. The most strident complaints about the imbalance come from the second and third tiers, who are either losing or likely to lose work as a result of the rapid increase in Japanese imports.

I reminded professor Cole of the Big Three's powerful in-house parts divisions, which supplied at least 65% of parts produced in the U.S. I said, "Japanese manufacturers are keen to buy more American parts, but cannot do so until the Big Three parts divisions, the pillars of the American parts industry, start selling their products to other automakers. By splitting up its parts division into a number of independent parts companies and allowing them to sell to others, GM will boost its restructuring efforts and contribute to the general growth of the auto parts industry."

Cole agreed with my suggestion but noted that labor and management relations made it all but impossible.

In August 1990, just after Robert Stempel's accession to GM's chairmanship, William E. Hoglund, the general manager of the Pontiac Division, was promoted to vice president and manager of the Automotive Components Group (ACG). Immediately he announced plans to boost ACG profits and productivity by selling parts to outside manufacturers as well as to GM itself. Sweeping staff changes were made in the area of parts counters. The parts purchasing manager at Toyota noted that, for the first time, "their attitude changed; whereas previously they always said no to whatever we asked, now they were actually starting to listen to us."

VI-8. The New Purchasing Philosophy

When Jack Smith became GM's president in April 1992, GM made a revolutionary change in its purchasing philosophy. Smith appointed J. Ignacio Lopez, vice president and purchasing manager for GM Europe (whom he had met at the Zaragosa plant in Spain while serving as chairman of GM Europe) to the additional post of vice president in charge of worldwide purchasing, giving him responsibility for purchasing at the NAO. Lopez immediately sent out a demand to all suppliers (including ACG, the in-house parts department) for quality improvements and price cuts of no less than 20%. He told them that henceforth there was to be absolutely no distinction between ACG and other suppliers. These moves created massive shock waves at the time but, backed fully by the new president, brought appreciable cost cuts in a remarkably short period. In March 1992, GM finance manager Charles Golden said, "Under Lopez we managed to cut $1.1 billion in 1992, and we're aiming for $2.2 billion in 1993."

The new vice president in charge of ACG, J. T. Battenberg III (appointed in November 1992 in conjunction with the replacement of Stempel with John G. Smale as chairman and the appointment of Smith to CEO), began cutting costs and working to boost the number of outside suppliers, in line with Lopez's new purchasing philosophy. While remaining within the company, ACG began to function as an independent supplier for the first time.

A research institute was set up in Japan to participate in "design-in" programs with Toyota and other Japanese manufacturers. ("Design-in" is a program under which an automaker and its subsidiaries undertake joint efforts to design designated parts or components from the very beginning of vehicle development.) Toyota steadily boosted purchases of American auto parts in line with a voluntary 1994 fiscal year target announced in January 1992; the total purchase value rose from $4.4 billion in FY 1992 to $4.7 billion in FY 1993, reaching $5.3 billion in FY 1994. ACG is now Toyota's major parts supplier in the U.S., while Toyota represents ACG's biggest outside customer.

In December 1992, GM said it would sell any part of that group that was not essential to its core business strength. It tried to sell 14% of its $25 billion business in automotive components. The sale of any part of the business or establishment of a fully independent outside company was conducted with the approval and cooperation of the UAW. (Under the new 1994 agreement with the UAW, workers at GM parts factories receive the same wages and fringe benefits — approximately $42 per hour — as their counterparts at assembly factories. The UAW agreement also specifies that any new arrangement reached with workers moved to the new company shall be valid for a period of three years.)

By August 1994, GM had sold off an Inland Fisher-Guide plant (manufacturing door handles, seat adjusters, and body moldings), three Delco Remy plants (electrical fittings and related parts), six Saginaw plants (propeller shafts and other parts for rear-wheel-drive vehicles), and four Delco Chassis plants (motors and actuators).

In March 1993, Lopez, the purchasing maverick of GM who had reduced purchasing costs of parts by 20% or more, resigned suddenly. Despite Smith's best efforts to persuade him to stay, Lopez went to the president of Volkswagen, Ferdinand Piech. Smith replaced him with executive vice president and CEO Richard Wagoner Jr., and announced that the parts cost reduction program would continue as planned. The vice president in charge of ACG, J.T. Battenberg III, was promoted to ACG president in June 1994. The *Wall Street Journal* of August 9, 1994 quotes Battenberg as saying:

"Annual ACG sales for 1994 will fall slightly, from $25.1 billion in 1993 to $25.0 billion. As mentioned earlier, factories that accounted for $3.5 billion in annual sales have been sold off, although sales have been boosted at the remaining plants. The number of employees has fallen in conjunction with the asset sales, from a national total of 111,600 at the end of 1992 to 93,800 by the end of this June. The asset sales will continue."

This illustrates how GM is working steadily to streamline the parts production structure. Although separating ACG as such is probably not achievable in the immediate future, GM clearly aims to convert it into one or more fully independent companies as part of the revitalization of the NAO. ACG has already been renamed Delphi Automotive Systems, and is represented throughout GM as an independent entity. Delphi president Battenberg (formerly president of ACG) expressed enthusiasm about the move toward independence during a press conference, citing as his model the Japanese company Denso, which gained independence from Toyota and now obtains 50% of its sales revenue from other companies. Similarly, Battenburg is keen to retain the NAO as his major client, while boosting the proportion of outside sales from 27.5% in 1994 to 30% by 1996, and to 50% by the year 2002.

This is one example of predictions by specialists that have never been realized. I experienced another example. While I was in the U.S., I advocated higher gasoline taxes as a means of reducing fuel consumption and alleviating or even eliminating the national deficit. Congress was then in the process of discussing a significant increase in CAFE standards to combat environmental problems. Some wanted a single numerical target of 40 mpg by the year 2005 for all manufacturers, while others argued for a uniform 20% increase by each manufacturer over the 1994 CAFE performance figures. This second proposal would clearly work against Japanese cars, which have much better CAFE standards already than their American counterparts. The second one currently seems more likely to be adopted, since it would serve as a protectionist measure favoring American cars.

In the U.S., large cars over 15 years old, so-called gas-guzzlers, account for 10% of all cars on the road because the price of fuel is so low. They are hampering the cleaning up of the environment. Higher gasoline taxes would force owners to switch over to more fuel-efficient cars. Arguing that simply fiddling with CAFE figures would not help the environment in the short term, I suggested a 25¢/gallon increase in the gasoline tax. This would induce people to save gas and at the same time would generate millions of dollars in revenue and help trim the deficit, killing two birds with one stone.

Republican congressmen discounted the idea as unfeasible. So did the Democrats and Maryann Keller. Professor Cole said, "He formed a research group on the matter, but in vain: recommendations for raising tax on gas submitted to the White House were firmly rejected by John H. Sununu, then Chief of Staff in the Bush Administration." My views were shared by Clifford M. Winston, senior fellow in the Economic Studies Program of the Brookings Institute. The Clinton Administration actually increased the gasoline tax shortly after coming to power, although on a very limited scale.

Under strong lobbying from automobile manufacturers, meanwhile, the government has shelved the CAFE question in favor of a joint research and development program with the industry for ultra-fuel-efficient vehicles.

VI-9. Rebuilding the Domestic Sales Structure

GM president Jack Smith had three strategies for sales: holding the program sales method in check, preserving the Oldsmobile division, and incorporating customer satisfaction into performance evaluation standards for dealers.

Program Sales

U.S. automakers made efforts to minimize the drop of domestic car sales share against the imported car penetration. They paid more attention to fleet sales, particularly to car rental companies, with the result that domestic fleet sales accounted for over 40% of total domestic car sales. The retail sales share of domestic cars plunged dramatically while imports, with small fleet sales, have remained high, at 45% in 1991.

Domestic fleet sales have increased as a result of the explosion in program car sales over the past few years. A program sale is an agreement under which the car rental company purchases vehicles on the understanding that the manufacturer will buy them back after a given period and at a special price. Both sides stand to benefit if the agreement is carried through properly. But John Rock, GM's vice president for Customer Sales and Service Staff (now general manager of the Oldsmobile division) said in the October 1991 issue of *Auto Rental News*, "Never forget that our aim is to maintain market share."

Automakers sell cars to rental companies, then repurchase them at a guaranteed rate and resell the units to car dealers at auction. These cars appeal to customers unwilling to pay the full price for a new car. For new car dealers, program cars mean customers and profits.

The Big Three used hard-pressure tactics in order to maintain sales, production volume, and market share, despite general market contraction due to the long-running recession and the threat from import penetrations. The Big Three competed among themselves even to reduce the fixed period of usage, and at one point the usage period was three months or less (normal terms were four months minimum). The overflow in the used-car market forced down their prices to well below expectations and caused massive losses for the automakers. On April 15, 1991, *Automotive News* called for an end to the scheme. According to the article "Program Cars Become a Drug: Now is the Time for Automakers to Say No," the Big Three were sustaining losses of $3,000 to $4,000 per vehicle in rental fleet sales.

As soon as he was made president, GM's Smith ordered fleet sales reduced because it was a direct cause of the operating deficit. The fleet sales' share of total car sales dwindled from 33.9% in 1991 to 30.6% in 1992 and 25.9% in 1993 (*Automotive News*, March 22, 1993 and March 21, 1994). In turn, the sales incentive cost per vehicle fell to $750 by the second quarter of 1994, significantly boosting profits at the NAO (*AP-DJ*, July 28, 1994).

Preserving the Oldsmobile Division

Until 1986, the Oldsmobile division had sold more than 1 million passenger vehicles a year, and was second only to the Chevrolet division in terms of sales. From 1987 on, the combination of the recession and encroaching Japanese imports had a drastic effect on Big Three sales. GM's Oldsmobile division was particularly hard hit; sales fell some 60% from 1986 to 1991 to barely 430,000 units, less than even the Buick and Pontiac divisions. The sharp drop in sales was attributed to the failure of the A- and G-body model change.

GM had also lost many customers due to chronic quality problems. I experienced these during a trip to the Canadian Rockies in 1988. Upon arriving at Calgary airport, I was asked to accept at 30% discount an Oldsmobile Ciera in place of the compact I had originally reserved. Over the course of a week I was amazed at the poor performance of the Oldsmobile: It ran badly, the chassis was poorly aligned, and the switch knobs came loose or fell off.

Market analysts outside the company had been calling for the Oldsmobile division to be scrapped since well before Smith became the president. Despite the fact that the G-body was too old and in need of a full model changeover, former president Reuss chose to retain the G-body and use it for developing the next model. The new executives under Smith also harbored initial doubts about this development plan. However, if GM dropped this car, not only the Oldsmobile division but GM would be bereft of completely new models until 1997. Abandoning the Oldsmobile division itself would mean a significant downsizing of the overall marketing structure and would be seen as tantamount to an admission by GM of its inability to keep the division afloat. This was too unpalatable, so the new car went ahead and the division was preserved. A joint letter from new chairman Smale and new CEO Smith was sent to inform dealers of the decision to retain the Oldsmobile sales network.

The new car was originally designed as a luxury vehicle to take the place of the Oldsmobile Toronado (the old front-engine front-drive model). It was named the Aurora and boasted a 32-valve V8 engine and optional extras such as a car navigation system. John Rock, the newly appointed general manager of the division, asked the approximately 3,100 franchise dealers to practice a "vision center" program, which was designed primarily to achieve the same customer satisfaction ratings as the popular Saturn dealerships. Dealers who implemented the program would be permitted to sell the Aurora. According to Rock, 1,700 of the franchises had undertaken the program by August 1994.

Production of the Aurora got off to a shaky start in June 1994. At one point, it seemed as if GM might not be able to produce 20,000 units in 1994, but output levels were restored to normal by autumn, exceeding 27,500 units. The new executives also decided to revive the Bravada, the only multipurpose sports car in the Oldsmobile division. The Bravada (captive import of the

Isuzu Trooper) was to be brought out as a new model during 1996. Oldsmobile dealers had to wait until 1997 for a medium-sized car like the Ford Taurus, Toyota Camry, or Honda Accord, which is reportedly to be developed on the same platform as the new Chevrolet Corsica.

These decisions, which illustrate the resolve of the new executive members, have every chance of meeting with success. Merging or abolishing sales channels is common practice among automakers seeking to rebuild. Ford abolished the Edsel division and Nissan its Cherry dealerships, while Chrysler merged Dodge with Plymouth dealers, and Mazda Enfini with Eunos dealerships. In every case, these moves reduced the overall selling capacity of the automakers.

The Customer Satisfaction Program

In a *Wall Street Journal* article titled "Buyer Satisfaction Survey Ranks GM Highest Among Domestic Automakers" (July 6, 1994), Oscar Suris reported the results of the most recent survey of car owners for the Customer Satisfaction Index compiled by J. D. Power and Associates. Led by the Saturn and Cadillac divisions, GM scored 136 points, compared with an industry average of 135. Ford and Chrysler were both below average.

Just after Jack Smith became president, GM made an effort to boost customer satisfaction, winning back lost customers and improving customer service at sales and service centers. Smith took the approach that had proven so successful with the newly established Saturn dealerships and developed it into a general program covering all dealership networks. The above-mentioned Customer Satisfaction Index demonstrated part of his success. According to *Automotive News* (June 6, 1994), GM is planning to incorporate the operating standards to be linked to its revised 1995 franchise agreement and eventually to be incorporated in a dealer's CSI rating. In June 1994, GM's dealers throughout the country were sent more than 100 operating standards, some of which are listed below.

GM Dealer Standards

- All salespeople must be able to accurately discuss their own and competitive products.
- The sales experience must not be high-pressure or confrontational.
- Salespeople must personally inspect every vehicle prior to delivery.
- Vehicle delivery should be a memorable event.

NAO Standards for GM Dealers

1) Customer assistance
   - A test drive will be offered to all customers.
   - All parties to purchase transaction will be treated as "decision-makers."
   - An orientation drive will be offered at delivery.
   - A member of management will thank all customers for purchase.
   - Dealers will initiate an owner-retention program by making follow-up contact after three days, five months, one year, and then annually throughout the customer's ownership of the vehicle.

2) Service
   - Dealers will provide service maintenance reminders to customers.
   - Alternate transportation will be available to service customers.
   - Vehicles being serviced will be fixed right the first time.
   - Serviced vehicles will be returned clean, with the radio turned off and electric settings restored as practicable.

Such standards are of course easy to draw up and demand, but harder to implement and maintain on a long-term basis. Nevertheless, GM has made a firm commitment to this approach.

The *Wall Street Journal* (June 6, 1994) reported that Toyota was the top company in the J.D. Power Customer Satisfaction Index with 150 points, and Lexus the top sales division with 176 points. Toyota is thus still way ahead, although if its customer satisfaction program is successful, GM could equal or even surpass Toyota.

The appointment of G. Richard Wagoner as president of the North American Operation in June 1994 was accompanied by the promotion of J. Michael Losh from marketing manager of the NAO to CFO. GM announced that the new marketing manager would be sought outside the company, and appointed Russell Reynolds Associates as a headhunter. The move, attributed to Smale, caused an enormous controversy both inside and outside the company, which had traditionally appointed officers from among its ranks. Inside GM, the emphasis had always been on the finance, technology, and production divisions, at the expense of the marketing division. The resulting lack of attention to marketing personnel obviously influenced Smale, an outside director.

The search for a replacement was unsuccessful for some time, prompting talk at one stage of appointing an insider after all. In November, a suitable candidate was finally found in 45-year-old Ronald L. Zarrella, president of Bausch & Lomb. On December 12, Zarrella's appointment as vice president and director of the North American Sales, Service, and Marketing Groups was publicly announced.

VI-10. Stepping Up Overseas Operations

GM's overseas operations, particularly in Europe, generated considerable revenue from the end of the 1980s through the early 1990s, compensating for deficits created by the NAO. (As the chairman of GM Europe, Smith pulled the company out of its operating slump under the lean production program.) From 1993 on, while Europe suffered a recession, production and sales in Central and South America soared. The first half of 1994 saw a new production record of 204,000 units in the region as a whole, excluding Mexico.

The Omega and Astra, developed by Opel in Germany, were incorporated into the product lineup of the NAO. The Corsa, which Opel brought out at the end of 1993, has proven extremely popular in markets throughout the world. In the U.S., its production will get under way at the Wilmington plant in Delaware during the 1997 model year. The Corsa is almost certain to play a key role in the expansion of overseas operations after 1995.

In June 1994, Louis R. Hughes was promoted from executive vice president to president of the overseas division. At the same time, Wagoner was appointed president of the NAO and Battenburg the president of the ACG, with all three reporting to the company president, Smith. Each division was given equal weight and expected to turn a profit and contribute to GM overall. Each is now steadily building up a presence in major markets including Japan, China, India, and Russia. The world's largest manufacturer, GM is set to become the world's largest, strongest, and best-known global organization not only in this century but on into the 21st century.

Chapter VII

Lean Production and the Challenge to the "World Car"

Lean Production and the Challenge to the "World Car"

VII-1. Ripple Effect of the MIT Study

In the fall of 1990, a book called *The Machine That Changed the World,* written by three scholars, Daniel T. Jones of Sussex University and James P. Womack (Ph.D in civil engineering) and Daniel Roos (Ph.D in political science) of MIT, was published by Rawson Associates. This book was the culmination of a five-year, $5-million study of the world's automotive industry conducted by MIT. This study was done under the direction of the International Motor Vehicle Program at MIT, which was established in early 1985, with the cooperation of Sussex University in England, and Harvard University, as well as the leading automakers of the world.

The book talks about how the Toyota production system — developed by honorary chairman Eiji Toyoda and former vice president Taiichi Ohno (in charge of production) during the 1950s — is far superior in terms of technical development, production, and distribution to the mass production system developed by Henry Ford in 1908. Furthermore, it goes on to say that unless American and European automakers switch over to this production system, they won't survive into the 21st century. The authors called Toyota's production system "lean production," versus Ford's mass production. "Lean" means "thin" or "without fat," but in this case refers to a system without waste. At the time, some argued that it should be called the Toyota production system, but I find the term "lean production" an excellent, appropriate way to express the ideas behind the system. Although it must be recognized that the general term "lean production" may be seen by some as different from the Toyota production system, this text nonetheless uses the two labels interchangeably.

Compared to GM's Framingham plant, it takes Toyota's Takaoka plant only half the time to build one car, and Toyota's defect rate for every 100 cars is one-third of GM's. In terms of parts stock, Toyota has only two hours' worth of stock, whereas GM has two weeks' worth. Regarding the space dedicated to assembly, Toyota needs only 60% of what is needed by GM. These productivity differences also appear in new model development; Japanese automakers that utilize the Toyota production system spend half the time required by American automakers for product development.

As a result of the two oil crises in 1973 and 1979, Japanese vehicle sales in the U.S. took a large leap because Japanese cars were small and fuel-efficient, whereas American cars were large gas-guzzlers. U.S. automakers claimed they could overcome this problem by focusing their efforts on making these types of small cars, thereby throwing the Japanese cars back into the Pacific Ocean. But this was not done. By the beginning of the 1980s, however, the majority of Americans acknowledged that Japanese cars were not only more fuel efficient but also had excellent quality, performance, advanced technology, style, and value for the money. On the other hand, the Big Three's small-car development program failed at every attempt or was far from successful, as mentioned in Chapter IV about GM. The Big Three had no choice but to admit that they could not build a high-performance small car of a quality comparable to that of Japanese cars and be able to offer it at a reasonable price.

Even during the early 1970s, it was commonly believed that the Japanese were able to supply low-cost, highly reliable cars because Japanese workers were skilled workaholics willing to work for very low wages. Many American business leaders believed that by making full use of America's highly advanced production technology, Japan's superiority could be eroded.

However, as the 1970s progressed, American car quality continued to worsen while Japanese car quality continued to improve, and the overwhelming superiority of Japanese cars became more apparent.

Consumer Reports each year publishes a special car issue that reports the statistical results of the evaluation of cars, new and used, by almost 1 million consumers. If you look through these special issues, you will see that the quality of Japanese cars improved dramatically from 1970 through 1980, surpassing that of American and German cars in 1980.

In the late 1970s, not a few American scholars and researchers (such as the late William Abernathy, a former professor at Harvard) began to study Japanese automobile production and management systems. Their reports shocked the top executives and managers of the Big Three, who could not believe the Japanese way of management could possibly be superior. The reports included the following observations:

1) In Toyota plants, anyone working on the assembly line can pull a cord to stop the line if he makes a mistake.
2) There is no repair area (or space, for that matter) in a Toyota factory at the end of the line. Vehicles that are lined off are shipped immediately; furthermore, the quality in which they are shipped is extremely high.
3) A die change in a Toyota factory takes only three minutes. In an American car plant, it takes 20 hours or more.
4) To increase efficiency and improve quality, there is an employee suggestion system. A great number of suggestions are received every month, and good suggestions are continually implemented.
5) To improve efficiency, work-in-progress parts stock is minimized. At the same time, to install quality management, the subsequent process takes from the preceding process only the amount of materials and parts that are needed, and at the time needed. (It was around this time that the now popular terms *kanban* or "Just-In-Time" systems were first introduced to the U.S.)
6) The absenteeism rate in Toyota plants is around 2% to 3%, compared to 6% in American plants.

After studying these points, some of the scholars and top executives of the industry concluded that it wasn't low labor costs that enabled the Japanese to produce high-quality cars at low cost.

Although the outstanding characteristics of the Japanese auto production system came to be widely known, some studies in the U.S. reported that the superiority of Japanese automakers was a result of cultural differences unique to Japanese companies and not applicable to the American environment. Specifically, the Japanese production system works because the Japanese have young, highly educated, deeply loyal workers who place their company first and their personal lives second. In turn, they are promised lifelong employment by their companies (*Productivity and Competitive Cost Advantage*, William J. Abernathy et al., U.S. Department of Transportation, February 13, 1981).

These comments, however, were not proved true at NUMMI, the joint venture between Toyota and GM, where production officially began in 1985 and utilized an American work force. Furthermore, 70% of the work force had belonged to the most notorious and radical branch of the UAW. Regardless, this group was put to work under the same production system Toyota had in Japan and was able to improve production efficiency and manufacturing quality, both of which were comparable to those of the plants in Japan. In other words, NUMMI proved that the Toyota production system was not something that would work only with the Japanese, but was a universal system that could be transplanted to the U.S.

The International Motor Vehicle Program at MIT studied the Japanese auto production system, concentrating on Toyota. It began the study by analyzing American and European auto production systems. Some of their conclusions:

1) Based on a substantial body of data, the study clearly showed a large productivity gap between Japanese automakers who had adopted lean production and American and European automakers who still utilized traditional mass production.
2) Japanese suppliers are not simply so-called job shops to the automakers. They have the capability to conduct the technical development of components and parts independently or jointly with the automaker. They also take responsibility for monitoring the quality of their parts, make the utmost effort to lower costs and increase productivity, and are required to make all efforts to deliver parts in time.
3) New product development is conducted in an extremely efficient manner. New model development in the U.S. takes on average 60 months but only 46 months in Japan. This is the fundamental reason the Japanese have such a strong product development capability. Other reasons are:
   - There is a project leader who is given the power to oversee all aspects of product development.
   - At the time the automaker embarks on a new product, suppliers are involved from the very beginning in the development and design of new model parts and components.

- The making of the body panel press die, which requires a great number of labor hours during production start-up, can progress in parallel with body design development.

The authors of *The Machine That Changed the World* state in their introduction:

> *We believe that the fundamental ideas of lean production are universal — applicable anywhere by anyone... We pay little attention to the special features of Japanese society, which some observers credit for Japanese success, but which we believe are of secondary importance...*
>
> *Lean production combines the best features of both craft production and mass production — the ability to reduce costs per unit and dramatically improve quality while at the same time providing an ever wider range of products and ever more challenging work... We believe lean production will supplant both mass production and the remaining outposts of craft production in all areas of industrial endeavor to become the standard global production system in the twenty-first century. That world will be a very different, and a much better, place.*

We were glad to see that someone had finally openly said that the Toyota production system was responsible for the superior quality, high performance, good styling, and competitive price of Japanese cars. Not only that, but this study also proved that this production system could be adopted elsewhere in the world. This work deserves high praise for demonstrating that somehow Toyota could contribute to the future growth of the world auto industry by establishing plants overseas, thus spreading the concept of lean production abroad.

The Shock Felt by European Automakers

It was the European automakers who felt the greatest shock when this study was made public, when the governments of Japan and the European Community (currently the European Union) were conducting negotiations regarding auto trade in anticipation of the 1992 unification of the European Market. But, with the revelation of how far behind the European automakers were compared to the Japanese in productivity and quality, it became clear that the Europeans would have to adopt this unfamiliar production system.

As a result, dissenting voices against liberalization were heard from top European executives such as Peugeot's Jacques Calvet, and they finally decided on a longer nine-year period to achieve liberalization. Not only did mass-volume European automakers such as Fiat and Renault suffer a big shock, but so did luxury automakers such as Mercedes-Benz. An article in *Fortune* (October 22, 1990) noted:

> *They discovered that one Japanese manufacturer can build a luxury car with 14 man-hours of labor — the same amount of time that an upscale European automaker spends fixing its expensive sedans at the end of the assembly process. The Japanese achieve identical levels of productivity and quality in luxury cars with one-sixth the European manpower. To obtain cooperation for their ground-breaking study, the authors promised anonymity to the 23 companies they studied. The most important statistical findings are grouped by continent so that differences among individual automakers are invisible. There is no way for the reader to learn, as* Fortune *did, that Toyota is the super-efficient Japanese luxury car-maker in the example above and Mercedes-Benz the labor-rich European one.*

Response by Ford Motor Company

Ford had continued joint venture negotiations with Toyota since 1979, but they broke off talks in 1981, as they were unable to reach an agreement. During this time, Toyota opened the doors of its plants and research facilities to the chairman and the president of Ford, as well as its top executives and managers. As a result, Ford was able to see the merits of the Toyota production system earlier than the other two Big Three companies. Ford was also helped by the cooperation received from Mazda, whose 25% of the stock Ford had acquired in 1978.

Donald Petersen, former chairman of Ford and the author of *A Better Idea* (Houghton Mifflin, 1991), talks in his book about the team system at Ford. This might be seen to reflect Ford's study of the Toyota production system.

According to the book *The Man Who Discovered Quality* by Andrea Gabot (Times Books, 1990), Petersen promoted reform within the company in order to ride out the recession of 1980. In doing this, he became a disciple of Dr. Demming, who played a decisive role in improving the quality of Japanese products, and convinced his reluctant subordinates to follow his teachings:

Ford gradually pieced together a strategy that would bring the company ever closer to the Japanese concept of total quality control (TQC). Top executives who represented a wide variety of disciplines and who in the go-go 1960s might well have thought twice before sharing a meal together began meeting regularly to set quality priorities and to establish standard corporate practices for making sure the needs of customers were met. And Ford operations began to work closely with suppliers.

Petersen's enthusiasm for quality control was so strong that almost a third of his autobiography was devoted to case study reports on total quality control (TQC). Ford made its own study of the Toyota production system before the MIT study and began to promote a shift to lean production at the beginning of the 1980s.

Response by General Motors

It was clear from his speech in Kyoto in February 1987 (see Chapter IV) that GM chairman Roger Smith hoped to spread the Toyota production system learned at NUMMI throughout the company. But the president at the time, Smith's direct subordinate, was not willing to support the introduction of the new production system. As a result, the fundamental measures needed for reform were not taken at the end of the 1970s or in the 1980s. GM's NAO (North American Operations) suffered huge losses, and GM's corporate constitution deteriorated to a level far below that of Ford and the Japanese makers.

For their study, the MIT survey team first visited GM's Framingham assembly plant in Massachusetts in 1986. In their report, they describe their visit as follows:

Our first interview with the plant's senior managers was not promising. They had just returned from a tour of the Toyota-GM joint venture plant (NUMMI) where John Krafcik, our assembly plant survey leader, formerly worked. One reported that a secret repair area and secret inventories had to exist behind the NUMMI plant, because he hadn't seen enough of either for a "real" plant. Another manager wondered what all the fuss was about. "They build cars like we do." A third warned that "all that NUMMI talk [about lean production] is not welcome around here."

It seems that this type of response was common at GM. In other words, there seemed to be no effort made internally at GM to make the most of what was learned at NUMMI.

Therefore, the publication of *The Machine That Changed the World* was quite a shock to the people at GM. As mentioned in Chapter IV, this report served as a positive stimulus and prompted reconsideration of lean production methods and a reevaluation of NUMMI. GM Europe's president, Jack Smith, the only top executive who understood lean production, took the lead in promoting reforms and had already achieved numerous results by the time the report was released. Today Smith is instituting corporate reforms as chairman and chief executive officer at GM.

Chrysler, under the strong leadership of chairman Lee Iaccoca, made efforts to eliminate waste and to reform its operation to a leaner system. It sent employees on long-term training stints to its partner, Mitsubishi in the U.S., or Honda's U.S. plants to learn about lean production.

VII-2. The Toyota Production System and the Ford Production System

Taiichi Ohno, the former vice president who established the Toyota production system, expressed the following views in his book, *Toyota Production System: Toward Effective Business Without Mass Production* (Productivity Press, 1988):

> *The Toyota production system and the Ford production system are both based on flow operations. The difference is that in the latter system much effort is put into the placement and management of the parts supplied to the assembly line, whereas in the case of Toyota the necessary parts arrive at the necessary time and in the necessary amount at the side of the assembly line in the final assembly process. In the U.S., parts of the same size and shape are made together; in other words, the traditional principle is making the lot large and mass producing parts in a planned manner to achieve the lowest cost. Toyota operates under the opposite principle: minimize the lot size to eliminate the unnecessary burden of having stock on hand, which results from making the lot size large...*

Under the methods established by Ford, the final assembly line flows smoothly, but no flows were created for the other processes. The successors of Henry Ford did not accurately understand Ford's intent — to create a flow for production — and ended up solidifying the concept "The larger the lot the better."

In the process of establishing the Toyota production system, I always kept in mind the Japanese market climate of "many types in small volume." As this is different from the American market characteristic of "few types in large volume," I thought it necessary to create a Japanese production system. In responding to market needs and producing in accordance with the Toyota production system, I thought that for this system, which had been refined under the Japanese market climate of many types and low volume, the condition of many types and large volume would be favorable. I want to emphasize that this is precisely why results are being demonstrated in the developed Japanese market. At the same time, I believe that this is why the Toyota production system is also valid in the U.S., which became a market of many types and large volume since the Sloan era...

The Ford production method did not take the operation flow beyond the final assembly line. The fact that Ford was able to respond to the shift from few types to many types with the same "few types-large volume" production plants inhibited progress. Ford and Toyota, however, are exactly the same in terms of the principle of creating a flow in the process.

In his article in the *Sloan Management Review* (Fall 1988) titled "Triumph of the Lean Production System," John F. Krafcik (B.S., Stanford University; M.S., MIT Sloan School; former NUMMI production engineer; researcher at the International Motor Vehicle Program; currently employed by Ford Motors) said:

Toyota adapted the large-scale, highly efficient, constant-flow production philosophy of the Ford Rouge complex to its small-scale facilities in an interesting way. Unable to achieve either the vast levels of integration or the standardized product volume Ford managed at this single complex, Toyota nevertheless paid homage to the concept of continuous flow by building a local network of adaptable suppliers and integrating them with assembly plants. But instead of building a standardized product like Ford's Model T, Toyota achieved the capability of flexibility producing a wide variety of products using continuous-flow principles, something Henry Ford would never have attempted. A manifestation of the capability was Toyota's ability to stamp sheet metal parts in lot sizes of a few hundred parts; doing so was possible because they had learned to change stamping dies in a matter of minutes instead of hours. This type of flexibility is still a great competitive advantage for those who have mastered it.

He also notes that upon careful examination of the characteristics of the Toyota production system, the true Ford method is more apparent at Toyota than it is at Ford. He adds, "This is no surprise. One of the founders of Toyota, former vice president Taiichi Ohno, says, 'If Henry Ford were alive today, he would have done the same as Toyota for his own production system.'"

I think that the MIT report overemphasized the differences between the two systems and that insufficient attention was given to the fact that both systems were based on flow operations and were not complete opposites. This may have generated misunderstanding. Instead of encouraging the incorporation of lean production, it may have caused some American politicians and part of the automotive industry as well as some Europeans to perceive lean production as a threat that should be eliminated. Those who believed in mass production would probably have reached the same conclusion if a flow production were adapted to carry parts and components horizontally to the final production line.

In other words, mass production plants are by all means able to shift to lean production. Perceiving lean production as a threat and turning the opposite way only makes mass production plants more antiquated. It is a waste of time and offers no advantage. It is best to respond with the belief that incorporating lean production means changing to the methods that were originally intended by Henry Ford, and that it only requires a decision to incorporate them.

VII-3. Competition and Production Leveling

In October 1991, I had the opportunity to visit Dr. James P. Womack, one of the authors of the MIT report, at his home in the suburbs of Boston and had a discussion with him for about two hours. Because we had limited time, I got right to the point at the beginning of our meeting. "Your book has now become the bible of lean production. You and your coauthors act as its preachers, and you have earned much respect. That is all fine and well, but unfortunately, I think there are two important points you have failed to mention in your book. One is regarding competition, and the second is about production leveling. Because you failed to mention these two subjects, even if one were to read the book from cover to cover again, one should not be able to master the concept of lean production."

He responded by insisting that in the American market, which always fluctuates plus or minus 25%, production leveling cannot be done. Hearing his response, I realized how difficult it was to fully comprehend the Toyota production system itself, although the authors had earnestly measured its effectiveness. Their book does not include any explanation of the Toyota production system or of lean production. Readers can understand the effectiveness of lean production but cannot learn a new production method.

It might be possible to learn lean production through other books, but is regrettable that this book lacks the two above-mentioned points when readers want to know the prerequisites for the introduction and promotion of lean production.

Competitiveness

Eiji Toyoda, honorary chairman of Toyota, once said, "It is not necessarily difficult to produce a car. The problem lies in how to produce it at the lowest cost." Taiichi Ohno also said the same, and stated that his utmost objective in developing the Toyota production system was to minimize every cost. And it was competition that drove Ohno and others to work toward this objective. Toyota's survival depended on whether it could surpass other automakers by producing superior cars at lower cost. This was the motivation that drove the company.

Production preparations for the Starlet, which succeeded the Publica, Toyota's first entry-level car introduced in 1960, was going on-schedule to meet the April 1973 introduction target. However, in the fall of 1972, a major problem erupted. It turned out that the production cost would exceed the planned cost by 70,000 yen if assembled at Daihatsu's Ryuo plant. There seemed to be no way to reduce the cost. When it came to production of entry-level cars, the largest problem always had to do with cost. This issue is inevitably a problem in any auto plant in the world, so it wasn't unusual that Toyota had the same problem. Because this was to be expected, there were those who said that they should halt production of cars that would not earn any profit.

At that time Nissan had a car to compete with the Corolla, but openly regretted the fact that it did not have an entry-level car like the Publica. Nissan believed that because it did not have a car to compete with the Publica, it could not catch up to Toyota, so it worked hard at developing a Publica competitor. Therefore, given the marketing strategy, halting the production of the car was a very difficult decision to make. Ohno had a strong conviction that the goal of any vehicle production engineer was to build a car for the public of the best quality at the lowest possible cost. Because Ohno knew that his teacher, Kiichiro Toyoda, the founder of Toyota Motor Company, had the wish to produce a car for the general masses, he finally said, "I will take responsibility for the cost reduction," and convinced the company. Compared to the Publica, which became obscure in the market after the introduction of the Corolla, the new Starlet soon surpassed 10,000 units per month in production and sales. Ohno commuted to the Ryuo plant four times a week after the commencement of production to personally instruct and oversee the personnel at Daihatsu. The goal was to teach them the Toyota production system to achieve cost reductions.

Although Toyota and Nissan have always fought a fierce battle for market shares in the Japanese market, the battle is symbolized best in the competition between the Toyota Crown and the Nissan Cedric. The two automakers competed fiercely on the following points:

1) The two cars not only adopted a fresh new design when their models changed but also improved in terms of engine performance and fuel efficiency. Each maker tried to best its rival at every model change.

2) The Crown, which had been outstanding until 1970, started to lose ground against the Cedric in 1972, 1973, and 1974. This happened because the style change in the 1971 model of the Crown was too progressive. Furthermore, in order to respond to abnormal growth just before the first oil shock, an unexpected production increase forced Toyota to sacrifice manufacturing precision, and major quality defects became apparent. Normally, model changes took place every four years, but to remedy this situation, the product cycle was shortened to three years. At the same time, production capacity was bolstered and quality boosted, reversing the relationship between Toyota and Nissan in 1975. The Crown's superiority was firmly established.

As these two examples demonstrate, the driving force behind the Toyota production system or the lean production system in areas of manufacturing and product development is the competitive spirit. Without competition, lean production would never have been developed.

In 1986, Toyota established Toyota Motor Manufacturing Canada (TMMC) in Cambridge, Ontario, and began production of the Toyota Corolla in November 1988. The J. D. Power Gold Plant Quality Award for producing the highest-quality car in all of North America was awarded to TMMC in July 1991, while Toyota Motor Manufacturing, USA (TMM), in Kentucky, placed first in the U.S. and second in North America. It was the competitive spirit of the Canadian workers at TMMC that drove them to beat TMM and build a higher-quality car than the American. (A similar thing can be said about TMMK in the U.S. The workers there had the fighting spirit to be better than their counterparts at the Tahara plant in Japan.)

Quality can be ensured the first time by making certain that each individual worker is aware of the fact that he or she must build a good car and not a defective one. A car, which is composed of over 10,000 parts, can be built only when a worker at each production point in the assembly line can determine whether something is defective or not. There is only so much one can do to correct defects upon inspection of the vehicle after it has been built. And there is no limit to what is considered quality. Regarding the Crown incident mentioned earlier, it is said that the reason for its failure was the decision to increase production based on an acceptable level of quality. Toyota learned how important it was to recognize that quality is something that one must continually strive to improve. And it is the competitive mind that makes each and every worker recognize this fact.

Production Leveling

Let me explain what is meant by production leveling. For example, say that the production volume schedule for three months is as follows: first month, 4,000; second month, 5,000; and third month, 6,000, for a total of 15,000. Of the 15,000, 12,000 are sedans and 3,000 are wagons. If there are 20 production days in the month, then 5,000 vehicles will be produced each month, or 250 vehicles per day. Each day, 200 sedans and 50 wagons will be produced, but this does not mean that sedans will be built in one lot and wagons in another. Rather, a production schedule is formulated for assembling one wagon for every four sedans built. With this type of production schedule, both in-house and outside suppliers can continue to produce according to plan, and can deliver parts "Just-In-Time" to the final assembly line. Also, because worker layout does not have to be changed every day, each worker's proficiency improves, thereby improving and stabilizing quality. This is called "production leveling" at Toyota.

But to ensure the stability of the production line in this manner requires a strong, capable sales division. First of all, it must be able to formulate a highly accurate sales forecast. Second, it must be able to motivate dealers to actually obtain results as forecasted. It must also have solid advertising strength and be able to enforce a substantial sales promotion policy. Third, it should be able to read market changes as quickly as possible, and react astutely to them. Lean production is not embodied just by the production side; rather, by combining its efforts with those of the sales division, a company-wide system can be established.

In October 1991, I visited Ford's Atlanta plant. I decided to go there because Kaneyoshi Kusunoki, former chairman of TMMK, had said to me, "Among the Big Three assembly plants, the Atlanta plant comes closest to a Toyota plant. It's an excellent plant that has achieved maximum efficiency through minimal investment. If you have a chance, you should definitely go there." The plant director, R. D. Anderson, differed from other plant directors of the Big Three plants who wore handsome business suits and never set foot outside of their splendid executive offices. Not only did he walk around the plant in a unique worker's uniform, but the plant had no executive parking or executive dining room. Absenteeism was the lowest in the U.S., 1%, and in terms of production efficiency, it was number one among all of Ford's plants.

When we started talking about *The Machine That Changed the World* by Daniel T. Jones, James P. Womack and Daniel Roos (Rawson Associates, 1990), I told him that this book fails to talk about production leveling. He agreed. He said that even if he wanted to level production, he couldn't do it at this plant because he would not be able to obtain the full cooperation of his salespeople. He said he envied Toyota, "which has a comprehensive companywide system that promotes lean production, including the sales division." At Ford, the salespeople would tell him that for this week they should build 50 wagons per day and next week 500 per day. Because production numbers fluctuate so widely, he said they must alter production plans every week. I was impressed that Anderson understood that production leveling is indispensable to lean production. He also said:

> *The automation rate at the Atlanta plant is the same as TMM, at 92%. But 50% of the automation is fixed by multi-welders enclosed in large frames (automation rate indicates the rate at which spot welding is done automatically at body assembly). At Toyota plants, one robot is assigned a single operation, making it easy to change operations. Since half of the welders are fixed at the Atlanta plant, it is not easy to level production to meet demand.*
>
> *It is difficult to produce right-hand drive vehicles for the Japanese market, for example. Also, since much labor and capital are required to prepare for a model change, Ford could only make a few exterior changes for the 1992 Taurus.*
>
> *Although Ford's Atlanta plant and the Toyota plant have the same automation rates, Toyota has flexible facilities which make production leveling readily possible, while the Atlanta plant has apparatus conditions that make production leveling difficult.*

An influential parts supplier who delivered parts to one of the Big Three companies went bankrupt in the summer of 1991. TMM also received parts from this supplier. It was already too late by the time TMM had learned that this supplier was facing bankruptcy, and the supplier went bankrupt as soon as the automaker closed the plant. In the opinion of some people at TMMK, if the Big Three plant had been leveling production, it could have prevented the supplier from going bankrupt. Although parts inventory had continued to increase due to poor sales, the plant had gone along without changing its production plan, and only after building a backlog of stock did it suddenly close the plant. Because of this, the cash flow from the automaker to the supplier stopped suddenly, and the supplier had to declare bankruptcy. In Toyota's case, if poor sales continued, the company would level down production and make every effort to avoid plant closure, and the supplier would also decrease its production accordingly. The supplier would also be able to continue supplying its customers, in this case, TMMK, thereby avoiding the danger of bankruptcy. As mentioned previously, reacting astutely to market changes is one of the strengths of production leveling.

To make the Toyota production system or lean production work, production leveling is indispensable.

VII-4. GM Europe and Lean Production

In December 1990, shortly after the MIT report was released, one of GM's top executives made a speech on lean production at the Harvard Club in New York. This stimulated my curiosity. In those days, people talked about how most of GM's executives had basically ignored the Toyota production system even though they could learn the merits of the lean production system at NUMMI, their joint venture with Toyota. But there was one executive among them who could talk about lean production, which was a brand-new term. I wanted to know what he had said. The speaker was GM vice chairman Jack Smith (also serving as president of GM Europe at the time). The occasion was the granting of the Belgian-American Friendship Award to GM, and Smith gave a speech titled "Global Competitiveness." Through one of my acquaintances I acquired a tape of his speech. The following is an outline of what he said:

The subject was the operation at GM Continental's Antwerp plant. Lean production had been implemented there two years earlier, and the plant achieved great results, to the extent that this plant was one of the largest, most modern, and most productive auto plants in the world. It had also implemented "Just-In-Time" materials management. Smith said,

> "As GM Continental is moving to become a lean producer, positive results are beginning to show. Quality is up, and production is running at an all-time high. These improvements are all the more impressive because they were achieved along with a radical change in work patterns and organizational concepts, and the launch of new car models... The leanest firms today are the Japanese, but lean production is not a Japanese exclusive. It is not a product of anything unique to the Japanese culture. GM Continental is working hard to prove it can be successfully implemented in Europe... in a European and Belgian way... through the active involvement of its work force, trade union leadership, and management."

GM Europe's recent activities have been quite remarkable. According to *European Motor Business* (The Economist Intelligence Unit, May 1991), GM Europe's results since 1987 have been as follows:

> *In 1989, GM's European operations accounted for 15.5% of GM's global sales turnover (including GMAC) but over 43.0% of GM's global net income. The figures for 1987 were 13.6% and 35.5 %, respectively. In 1990, GM's European operations accounted for 19.6% of GM's global sales turnover and made a net income of $1.9146 billion compared with a GM global loss of $1.9857 billion (including a $2.1 billion after-tax charge for plant closing).*
>
> *Similarly, in 1991, GM as a whole suffered deficits, while GM Europe continued to make profits from the previous year.*

For someone like myself, who had focused his studies on GM's activities in North America, it was surprising to find that GM Europe was utilizing lean production and doing so well. It's also a great relief that at least one top executive at GM has close to a complete understanding of lean production. As mentioned in Chapter IV, having served as the main negotiator in the NUMMI joint venture talks with Toyota, Smith had acquired a good understanding of lean production. What makes him an extraordinary manager is that he immediately implements something when he thinks it is good. Lean production was introduced into GM Europe, where he was in charge, in the twinkling of an eye. The results previously mentioned followed.

Aiming to completely transfer the lean production system into Europe through the efforts of GM Europe alone, he promoted the construction of new plants. Germany's Opel invested an enormous amount, approximately $660 million, to completely renovate the Wartburg passenger car plant in Eisenach in the former East Germany and began production of the Astra in 1992.

The *Financial Times* made the following report on the plant (April 1, 1992):

> *At the old Wartburg plant, it took 9,000 workers to produce 70,000 passenger cars per year, while Opel plans to produce 150,000 with 2,000 workers. President LaSorda makes it clear that he is aiming to construct a high productivity plant that measures up to Toyota when he says he is aiming to reduce the number of man-hours it takes for assembly at Europe's passenger car plants (from 36 hours) to 20 hours or less at Eisenach. Japanese plants take 13 hours, North American transplants take 18 to 19 hours, American plants take 25 hours, and European plants take 36 hours. This plant has no stamping plant and operations resemble those at Canada's TMMC, but the production scale is estimated at approximately 50 percent more than that of TMMC.*

All employees, including the top manager, wear uniforms. The plant is cleaned in the same manner as at Toyota, by employing an outside cleaning company. Like at TMMC in Canada, seats are assembled by an outside supplier and delivered to the line when necessary. Various efforts have been made throughout the plant to incorporate the "Just-In-Time" system and other aspects of the Toyota production system. Productivity has increased steadily, reaching 4,300 vehicles in 1992 and 54,500 in 1993 (according to *European Motor Business*, Fall 1994). Since 1993, the plant has been producing the new Monza. It is now said that the plant is reaching the same level of productivity as that of the Nissan's Thunderland plant in England, the most efficient one in Europe.

Opel's president at the time was Louis R. Hughes, who is currently the head of the overseas division. With experience working at NUMMI, Hughes has a good understanding of the Toyota production system. In February 1991, Opel's chairman and vice chairman, together with labor representatives, visited TMMC. TMMC, an outstanding plant that has achieved high production efficiency with minimal investment, was pronounced the number-one plant in North America by J. D. Power in its 1991 quality survey. The Opel executives' enthusiasm for lean production was very apparent.

As for the NAO of GM, although it had a good, readily accessible model called NUMMI, it failed to learn from it and implement lean production. With the publication of *The Machine That Changed the World*, GM realized for the first time the significance of lean production. It can be said that this book was written for the benefit of those at GM's NAO, and for some European automakers who were also slow to wake up.

One executive in GM's North American operations summed this all up to me during my visit to the U.S. in the fall of 1991: "We have completely wasted the past 10 years. We were too late. We probably won't be able to catch up to Toyota even in 10 years." (Later he asked me not to quote his name. Perhaps the MIT report, too, was 10 years too late.)

However, the new president, Smith, is working to bring the European success to North America and introduce lean production to all plants by 1996. In February 1993, LaSorda, the president of the Opel Eisenach plant, which succeeded in introducing lean production, was assigned to the post of manufacturing manager for the Cadillac/Luxury Car Engineering and Manufacturing Division of GM's North American operations (he was succeeded by Eric Steven, who had management experience at CAMI). He was put in charge of GM's core plants of Hamtramck in Detroit, Buick City in Flint, and Orion Township. This transfer of LaSorda can be seen as part of the strategy to introduce lean production into North America.

VII-5. World Car's Conditions

GM developed the J-Car under the "world car" concept, but then suffered a setback, as mentioned in Chapter V. Research and study on this concept were continued, however, not only by GM but also by Ford and European automakers.

The Corsa, developed for the European market and put on sale in 1992, began to achieve world car status as mentioned before, winning acclaim not only in Europe but also in Brazil, Mexico, Thailand, India, and other countries. Preparations for global production are under way, with production in Brazil having already started. In fact, Opel has also started laying the groundwork for production of its Astra and Vectra, the larger predecessors of the Corsa, beyond the Western European region and into Poland, Hungary, and other countries in Eastern Europe. Similar preparations have extended into Turkey, India, Thailand, Indonesia, and Brazil, with some of these countries already launching production. The world production of the Astra in 1994 is estimated to have reached 700,000 units.

While GM failed to develop a world car of North American origin, Opel was unexpectedly successful. To improve the marketability of its products, the NAO is exploring the possibility of developing Cadillac and Buick on the same platform as the Opel Omega. Introduction of the Opel platform is allegedly being considered even by the Saturn division, which has traditionally pursued its own path, drawing a clear line between itself and the NAO.

At Ford, production and sales of the Contour and Mercury Mystique were launched in the U.S. in 1994. These cars were the U.S. versions of the Mondeo, which was developed by Ford of Germany and put on the European market in 1993. Ford had production startup problems, and initial sales were not as good as expected. Trottman, the new chairman of Ford, announced the global integration of Ford's development structure in the fall of 1994 and is currently linking the operations in Detroit with those in Europe. While the practical effects of this move are expected to emerge around the year 2000, it remains unclear how Mazda, which has contributed enormously to the development of vehicles produced in North America, will be incorporated into this structure. It is expected that Ford will rely on the technology of

Mazda in the future. The principal aim behind this new structure seems to be the development of a world car.

In October 1994, Volkswagen's chairman Piech spoke to journalists at the Paris Motor Show:

> "VW is renouncing its policy of selling the same car throughout the world beginning with the next generation of cars. The idea of the same car being adequate emerged as a result of VW not keeping in touch with consumer interests. This attitude became responsible for the drop in sales for the VW and Audi in the U.S. following a peak period in the 1970s. For the American market, we're going to make a completely different version (some suggest this will be a small front-wheel drive reincarnation of the old Beetle) of the Golf and Jetta, which are produced in Germany. To achieve this, there will be an increase in the types of car body from 20 to 30, but a decrease in platforms from 16 to 4."

Piech then announced VW's plans to shift to the so-called Plateau 8 production system, which would make it possible to assemble a new city car (using the same platform as the Polo, announced at the Paris Show) in only seven hours (*Automotive News*, Oct. 10, 1994).

Note: J. Ignacio Lopez, who moved to VW to become vice president under Piech despite being offered the post of president at GM's NAO, had announced his idea that breaking the production methods at assembly plants into several more highly advanced stages would make a more efficient system (what he referred to as "PICOS"), and had been proposing the construction of a new high-level, high-efficiency plant in his native land in the Basque region of Spain, ever since his days with GM Europe. But as GM executives were emphasizing the market in Eastern Europe at the time, his proposal was not accepted. The rumor is that Piech succeeded in pulling Lopez away by approving the construction of this plant. The Plateau 8 system announced by Piech refers to the aim of having the main components assembled and delivered by suppliers and holding the final assembly-line man-hours to approximately seven. Lopez is currently in a court battle with GM over charges that he took out confidential GM documents when switching over to VW.

In 1979, GM president Estes defined "world car" as "a car with the same basic design and many common or compatible parts, which can be adapted to market demands with minimal revisions, produced at a low cost due to globally integrated operations, and compete in major markets." (See Chapter V.) However, it can be said that the concept of the world car has changed with car design, production technology advances, and changes in consumer lifestyles over the last 15 years. Looking at the latest trends of GM, Ford, and VW, the following definition seems more appropriate today: "A multiple of cars appropriately priced to sufficiently meet user demands for lifestyles across various income groups, while using a minimal number of platforms composed of combinations of several engines, transaxles, and other main components and meeting the regional taste of each market of the world."

In short, the era of carmakers pushing a single model that they find most convenient for themselves has ended, giving way to a new era demanding a global production network that provides, at the lowest possible costs, a multiple of cars well matched to the various user needs in each region. It seems that cars produced under this kind of global production structure will become "world cars"—a direction, if you will, that combines the concept of VW chairman Piech previously mentioned and the actual trends at GM and Ford.

In China, the government has been in the process of selecting carmakers to undertake large-scale mass production of passenger cars since last year. The world's major carmakers are rushing in, but most of them are proposing next-generation entry-level cars in order to get a foothold in the Chinese market. The thinking is that cars that succeed in China, the world's largest market, automatically qualify as world cars.

It seems a vain effort, however, to argue about what an entry-level car should be in China, with its billion people leading diverse lives. China will one day have more cars than anywhere else in the world and consume more fossil fuels. Bearing this in mind, the only sure thing is that the deep, wide continent of China should have clean cars with high fuel efficiency. In 1972, I went to Brazil to conduct a market survey and was shocked at the terrible smog where I had imagined blue skies and green jungle in Saõ Paulo and the surrounding area. At the time, the local VW plant had succeeded in mass producing and marketing the old Beetle, and the region was in the midst of a major car boom. The town, however, reeked of car exhaust, and smog filled the sky. In China, significant smog has already emerged in some of the cities along the Yangtze River. Hopefully, the mistakes of Brazil will not be repeated. I also think, however, that there should be a wide variety of cars, as long as they fulfill the conditions related to clean air and environment. Based on our experience, it can be expected that freer competition will surely develop a stronger car industry in China more quickly.

Epilogue

Epilogue

The Arrival of the Black Ships

For the Japanese automobile industry, the year 1994 was very similar to the year the fleet of "black ships" came to Japan 140 years ago. 1994's sudden flood of American vehicles in Japan caused shock waves much like those generated when the fleet of U.S. Navy ships (called "black ships" because they were painted black) commanded by Admiral Perry arrived at Uraga Bay near Yokohama in 1854. The admiral asked the Tokugawa shogunate government to suspend the national isolation policy and forced it to sign trade and port treaties with the result that Japan opened up its market to foreign trade.

Before 1994, imported vehicles had mostly been expensive models from Germany and other European countries, which have posed little threat to Japanese cars. The scale of vehicle import had been similar to the trickle of foreign goods through the port of Nagasaki (the only port in Japan open for limited foreign trade) during the shogunate era.

Even after the import of vehicles was liberalized in Japan a quarter of a century ago, U.S. automakers did not try aggressively to enter the Japanese market. As a result, that market has remained effectively isolated, as in the days of the shogunate government.

As Chart 6 shows, before 1983, imports accounted for barely 1% of the market overall (excluding minivehicles), rising above 2% in 1987 and then 3% in 1989, and finally topping 4% in 1993. Most of the increase, however, represents "reverse imports" from Japanese transplants in North America. This trend never posed any significant threat to the domestic industry.

In 1994, however, total import volume topped 300,000 units, a 50% increase over the previous year, and the share of import vehicles in the Japanese market shot up to more than 6% of the total registration. Annual imports from the Big Three climbed spectacularly by 80% to reach 46,000 units, with passenger cars up 50% to 26,000 units and commercial vehicles up 90% to 10,900 units. Including imports from German subsidiaries of GM and Ford, total sales of American-brand vehicles topped 66,000 units, an increase of 60%. Similarly, British and Swedish vehicles both rose 30%, and German vehicles 20%. Reverse imports from Japanese transplants in the U.S. and other countries rose 130% to 84,000 units, or 28% of all vehicle imports.

This triggered an intense battle between domestic and imported vehicles in the Japanese market. Imports were not being sold uniformly throughout the country; instead, the vast bulk of sales was concentrated in the Tokyo metropolitan area (Tokyo and the three surrounding prefectures of Kanagawa, Chiba, and Saitama). Some 100,000 imported cars (nearly 40% of the total) were sold in the metropolitan area. In Tokyo, the world's largest car dealer, Tokyo Toyopet Motor Sales, has more than 80 outlets and 5,000 employees. But the sales of imports in the area surpassed even this mammoth dealer's sales volume, suggesting that the combined sales and marketing ability of import dealers is now greater than that of any one domestic dealer. The import car share is 8.1% nationwide (excluding mini-cars), but 17% within the 23 wards of the Tokyo metropolitan district, rising even further in certain wards such as Shibuya, Setagaya, Meguro, and Minato, where import share is 33%, 26%, 26%, and 20% respectively.

This means that one in every three cars sold in Shibuya, one in four in Setagaya and Meguro, and one in five in Minato is now an import. Residents there are in high-income brackets and are lifestyle leaders in Japan, suggesting that imports could rise much further in the future. And since these areas represent an important constituency of domestic car dealership in Tokyo, the rising popularity of imports could well make it difficult to maintain domestic market share as well as sales volume. Indeed, it is no exaggeration to say that the threat has struck at the very foundations of domestic car dealers, destroying their comforting "isolationist" scenario and signaling the onset of a battle for survival.

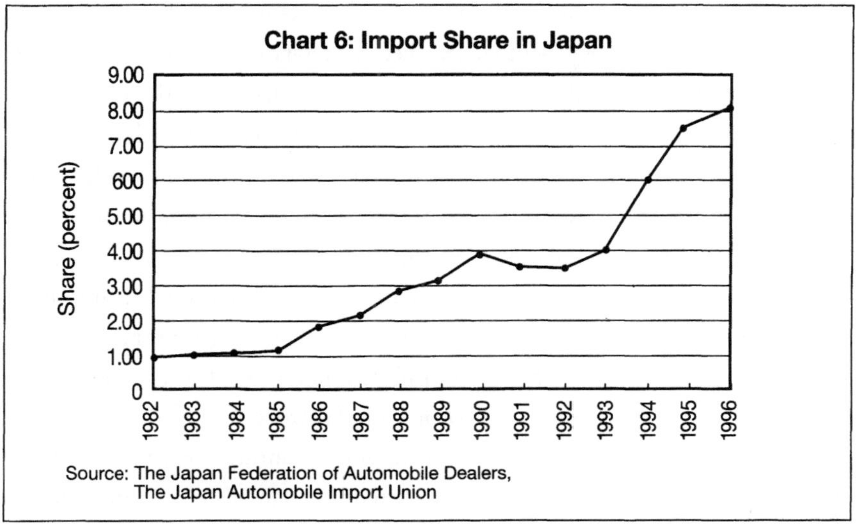

Both public and private sectors in the U.S. have called for an urgent response to the trade deficit with Japan since 1990, when the automobile and auto parts trade accounted for two-thirds of the overall deficit. The argument runs as follows:

Japanese transplants in the U.S. import most of their main parts from Japan. Even those parts they purchase in the U.S. come from affiliated Japanese parts manufacturers who set up operation at the same time their parent automakers built plants here. Many of these parts manufacturers are little more than "screwdriver assembly shops" relying on CKD production of imported Japanese parts. A rise of production at Japanese transplants simply means an increase in parts import from Japan. Thus, the import of finished vehicles is gradually declining while imports of parts is rising. This is why the deficit in automobiles and auto parts shows no sign of decreasing. To reduce the trade deficit, Japanese automakers in the U.S. should buy more from longstanding local parts manufacturers (traditional parts makers), and vehicles manufactured in Japan should use more American parts.

Harold Poling, then Ford's chairman (1990-95), claimed that it was abnormal that imports had only less than 5% of the Japanese market (conveniently ignoring the lack of effort in boosting exports to Japan). He had also sought the opening of the sales network to handle imports, and declared that

Japanese automakers should decrease their exports to the U.S. by 20% each year. He supported Robert Mosbacher, then Secretary of Commerce, in putting the U.S.-Japan automobile trade issue on the main agenda at the U.S.-Japan summit meeting in January 1992 held in Tokyo between President George Bush and Prime Minister Kiichi Miyazawa. Poling accompanied President Bush together with Robert Stempel, then GM's chairman, and Lee Iacocca, then Chrysler's chairman. They met with top executives of Japanese automakers and representatives of their dealers. (Stempel was originally reluctant about Poling's plan, taking GM's traditional stance of advocating free trade, but was eventually swept along by the tide of public opinion. He was also hesitant to visit Japan, but was advised to go by outside directors. As described earlier in Chapter VI, Stempel's downfall was being plotted during this trip.) The meeting achieved nothing. The Americans argued that Japan's closed market was hindering their exports, while the Japanese countered that their market was quite open and that the American automakers were simply not trying hard enough. The two sides remained at loggerheads.

Concerned about the potential breakdown in government negotiations, MITI asked the Japanese automakers to make their own voluntary plans for purchasing auto parts made in the U.S. and announce their plans individually. The total reached $19 billion. The figure was nothing more than a total of each manufacturer's plan, but the U.S. government tried to interpret the $19 billion as a numerical target, demanded further increases, and monitored progress toward the target. Toyota managed to fulfill its own voluntary purchasing plan, but the $19 billion proved difficult to attain, due to the rising yen.

As soon as Bill Clinton won the November 1992 U.S. presidential election, he started to build a very close relationship with the automobile industry. During the campaign Clinton had pushed strongly for marked improvements in fuel economy as one of his environmental strategies. The Big Three asked him to scale down his figures because they would be unachievable, due to the scarcity of research and development engineers and the heavy burden of investment required.

Clinton had also promised a strong stance toward Japan, setting the numerical target in order to eliminate the trade deficit. This was consistent with the aims of the automobile industry, which tried to maintain close ties with the administration. As the UAW was one of Clinton's most important constituents, his administration had to view positively the revitalization of the auto industry and the preservation of jobs, which was the prime concern of the UAW. This also served to strengthen the government/industry relationship.

Note: Clinton wanted CAFE (Corporate Average Fuel Economy) to be improved 40% from the existing 27.5 mpg to 40 mpg by 2001, and then to 45 mpg by 2006.

In September 1993, the U.S. government announced a joint 10-year $1 billion project with the industry to develop a "supercar" capable of 80 mpg (three times as much as the current performance level), and simultaneously shelved the planned tightening of fuel economy regulations. At his first meeting with Prime Minister Miyazawa in April 1993, Clinton managed to include automobiles and auto parts in the sector-specific U.S.-Japan Framework Talks. At the second meeting in July, it was agreed that objective criteria would be discussed for measuring the efforts made by both sides. During the ensuing negotiations, the U.S. government asked Japan to adopt numerical targets as objective criteria, with occasional threats such as the revival of Section 301 of the U.S. Trade Act. The Japanese government, wary of the way in which the 1992 voluntary plans were unilaterally treated as numerical targets, opposed it. Agreement still had not been reached by the deadline of September 30, 1994. Negotiations resumed in late March 1995, and continued for many weeks thereafter.

Although vehicle imports into Japan were fully liberalized in 1965, until the 1990s, American automobile manufacturers made no attempt to develop and produce vehicles tailored to the Japanese market.

Yanase Motors has been an import distributor in Japan for 77 years, being the importer of GM cars since the end of the World War II. Jiro Yanase, whose father started the business, is the current chairman of the company. In an article titled, "It's the Big Three Who Is at Fault—Why Don't They Build Something They Can Sell in Japan?" in *Bungei Shunju* magazine (March 1992), he said:

American and European automakers have habitually insisted that Japan should buy more of their cars, but regarding the current American issue, these automakers should be criticized appropriately, because in most cases they have not fully researched the Japanese market and provided a car that Japanese would want to buy on their own. It's like the example where a Japanese customer is forced to buy cabbage, although he really wants to eat sardines. This is what happens in many cases.

The main reason U.S. automakers didn't move into the Japanese market was supposed to be that they simply didn't have any models that could match Japanese cars in terms of quality, performance, or price, and weren't prepared to invest the time and money required to develop such vehicles. From the very beginning of the 1990s, their obsession with Japanese cars suddenly metamorphosed to a superiority complex.

In those days, due to the strong yen, American vehicles gained advantages over Japanese in price in the domestic market. Even in the Japanese market they became competitive in price. The companies started to campaign that their efforts to improve quality—their most vulnerable point—had paid off. They demonstrated the quality of their cars to consumers, politicians, and even themselves. Conveniently ignoring their past lack of effort in the Japanese market, they criticized Japan for its allegedly closed market, an unfair trade barrier that obstructs imports of their competitively priced, high-quality vehicles. Naturally enough, Japan has not agreed with them.

Note: The forceful stance of Clinton's administration can be attributed to the influence of so-called revisionists in his cabinet. His top economic advisor, Dr. Laura Tyson, was said to have become a revisionist under the influence of Clyde Prestowitz, director of the Institute for Economic Strategy in Washington, D.C. Dr. Tyson has advocated numerical targets in trade with Japan. She believed that even if Japan liberalized its trade restrictions, the trade imbalance would not be reduced because Japan has a fundamentally different brand of capitalism from the West. In a report, *The Future of the American Automobile Industry* (see Chapter VII), Prestowitz gives very detailed advice on how to address the automobile dispute between Japan and the U.S. This report represents the origin of the Labeling Act (enforced October 1, 1994), which requires every vehicle sold in the U.S. to bear a label indicating the percentage of parts from North America. (This act discriminates against foreign-brand vehicles built by overseas companies at factories in the U.S. because the local content figure in the label allegedly ignores the values and efforts of American workers.) Many recommendations in the report formed the basis of the U.S. stance at the Japan-U.S. automobile talks. In December 1993, Prestowitz released another report claiming that Japanese manufacturers had effectively put 135,306 U.S. workers out of work, since more workers had been laid off by local manufacturers than those who had found jobs at Japanese transplants. In this way, Prestowitz has played a central backstage role in Japan-bashing during the process of automobile trade negotiations. U.S. trade representative Mickey Kantor said that he maintained close contact with Prestowitz around the clock.

Ford's chairman Poling, an advocate of politicizing the automobile trade issue, criticized the closed nature of the Japanese market. On the other hand, he took positive actions to build up a sales network in Japan. In February 1991, Konen Suzuki, a former general manager at Toyota, was appointed president of Ford of Japan. Suzuki had left Toyota in 1988 to take up the position of executive director of the Toyota Nagano Motor Sales Co., the top dealer in Nagano prefecture, at their strong request. In 1990, he was contacted by an American headhunting firm. He told me later,

> "I knew that the number of candidates was being whittled down during the selection, and it made me more determined to win. I wrote a thesis with all my might. That got me a meeting with Vice President W. Waven Booker [former General Manager for Asia-Pacific Operations, later executive vice president for international operations]. I was impressed by his dignity, grace, and executive bearing, and despite being a man of few words, he convinced me to be hired by Ford..."

A number of other Toyota people have also switched to Ford. One of them is Yoshikazu Nakajima, the general manager of the Vehicles Division. He worked for me for a few years, then was sent to Columbia University in New York under the company's overseas training program, where he was awarded the MBA after two years' study. He returned to Toyota only to leave shortly afterward due to disputes after the merger of the factory (Toyota Motor) and the sales company (Toyota Motor Sales). It is said that he entered Ford of Japan at the behest of Suzuki. Those at Toyota who know both well say that they represent a powerful combination that Ford of Japan needs.

Since 1978, when Ford acquired a 25% share in Mazda through the Sumitomo Bank, the two automobile companies have been working closely on vehicle development and production in Japan, the U.S., Mexico, and elsewhere. In 1989, the Ford Asia-Pacific headquarters moved from Australia to Tokyo. Vice president Booker, based in Tokyo, saw Japan differently than his counterparts in Detroit; the *Wall Street Journal* (November 19, 1991) quoted him as saying, "The leading global automobile companies have to establish a presence in the Asia-Pacific region if they want to remain so. If the Big Three will mount a serious, long-term effort and not expect instant profits, the Japanese market is open to American cars."

Under Booker, Mazda started producing Ford vehicles in 1992 on consignment, to be sold through Autorama, set up by Mazda. (Ford contributed 34% of the funds and Mazda 39%, with the remainder being solicited from independent investors. In June 1992, Ford increased its share to 36.5%, the same as Mazda, and also appointed Suzuki chairman and president. Ford effectively started to control the operation.)

Ford regained control of all importing and retailing rights in Japan for its vehicles, formerly held by several importers including Mazda. It also began establishing its own direct retailing structure within Japan.

In 1991, sales of Ford vehicles (most of them manufactured by Mazda) topped 76,000 units, making Ford the leading foreign manufacturer in Japan.

Note: Since sales figures in Japan are compiled by the manufacturer, Ford vehicles produced by Mazda are counted as Mazda sales. It is thus not possible to determine total Ford sales from the official statistics. The figure above was supplied by Ford of Japan. In the U.S., on the other hand, sales figures are compiled according to the manufacturer name used for retailing purposes; thus, the GEO Prizm made by NUMMI is counted as a GM car, and the Corolla made by NUMMI as a Toyota. This applies also to imported vehicles relabeled as local brands ("captive imports"). Sales by overseas manufacturers (including sales of vehicles made in the U.S.) are treated as imports. Vehicles from U.S. manufacturers are classified as locally made, irrespective of actual origin. This makes the proportion of imports appear slightly lower and that of locally made vehicles slightly higher.

With the burst of the bubble economy in Japan, sales plummeted to approximately 60% of the previously mentioned level, but imports of U.S.-made vehicles such as the Taurus and Explorer began to rise again during 1993. This was followed by the new Mustang and German Mondeo in 1994, when annual sales of Ford imports reached 15,000 units, an increase of 170% over the previous year. Languishing sales of Mazda-produced vehicles finally began to turn around, reaching an annual 32,000 units.

By the year 2000, Ford plans to sell 200,000 vehicles annually, half imported and half manufactured by Mazda. Ford of Japan estimates that this will require some 700 outlets. Since it currently has just under 300, the key to achieving the target is opening enough new dealerships in the next five years. Suzuki knows very well from his time at Toyota that it is not just the number but the quality of the dealers that matters. For this reason he is trying to attract interests from Toyota and Nissan dealers, the most powerful in Japan. Major Nissan dealers have already opened Ford outlets in Tokyo, Kanagawa, and Osaka, and these are performing well.

Ford has also approached Toyota's dealers in Chiba, Kanagawa, and Aichi prefectures. According to a report in the *Nikkei Sangyo Shimbun* (March 29, 1994), the president of a Chiba dealer was interested in handling Ford vehicles. When he sought approval from a representative of Toyota, he was told yes but decided not to go ahead because "I could see in his eyes that the answer was really no." The story was immediately relayed to Washington and provoked a stinging reaction from the U.S. government, which saw it as further evidence of the closed and unfair nature of Japanese markets. (A reference even made it into the 1994 *Annual Report on Trade Barriers in Foreign Countries*, released by the office of the U.S. trade representative just days later on March 31. The section in the report on the automobile dispute between Japan and the U. S. stated, "The closed relations between dealers and manufacturers hinder access by foreign manufacturers to the dealership network.")

As it turned out, the dealer in question went on to announce in October a contract with Ford to open up three new Ford outlets.

Toyota had the same experience when trying to build up a dealership in North America. The owner of a leading Big Three dealership was summoned to Detroit after he had negotiated with Toyota and been threatened with dismissal. But he rejected the automaker's demand by telling them that the decision on whether he would sell Toyota or not was his responsibility, and he opened a Toyota outlet as planned. Sales managers of automakers are desperate to move as many units as possible, always pushing the outlets under their jurisdiction to sell one more unit. Thus, in this sense, it was perfectly natural for the Toyota representative to receive the dealer with a verbal "yes" accompanied by a facial "no." It is unfortunate that the dealer put the blame on others because he could not make the decision himself.

It should be remembered that the word "dealer" differs in Japan and the U.S. In American industry, a dealer is normally the owner of a retail outlet, while in Japan a dealer is a retail sales company as a whole. Neither side of the auto trade talks has attempted to clarify the difference of meaning, with the result that their arguments became somewhat confused.

A number of owners of Toyota dealerships actually hold more than one franchise besides Toyota. In the U.S. many owners hold more than one franchise, but a sales outlet never sells both GM and Ford vehicles. One owner recently opened a massive automobile dealers' park where several outlets under different franchises are in operation. Even in his case, the park is divided up according to the brand name of each franchise.

As I previously mentioned, no GM or Ford dealer (in the American sense of the term) showed the slightest interest in Toyota when we started to establish a presence in the U.S. It was a long time before Toyota began to make a name for itself and people started to inquire about franchises. It should be recognized, then, how futile and unreasonable it is to blame the Japanese government or Japanese automakers for the present situation in terms of the dealer-manufacturer relationship.

The rapid increase in American vehicle imports during 1994 was not caused by Ford alone. The advent in 1993 of the right-hand-drive SUVs (sport utility vehicle) Jeep Cherokee and Grand Cherokee boosted Chrysler sales by 130% to 13,000 units in 1994. Prior to this, during the disputes in the automobile sector of the Japan-U.S. Framework Talks, U.S. negotiators had tried to force their Japanese counterparts to provide numerical targets for sales if Chrysler built the right-hand-drive SUV for Japanese market. Honda had tried to sell Chrysler products once, but without much success, due to their high price and the left-hand-drive configuration. The introduction of a right-hand-drive model at a lower price, however, coupled with the continuing SUV boom, triggered explosive sales growth.

Both automatic and manual transmissions, which represent one of the main components of the Cherokee, are of Japanese make. Chrysler has quality problems with many of their products but few with these, contributing to strong sales in Japan. Chrysler hopes to sell 100,000 units per year, including the Neon, a small, cheap model that will appear next year, and the Cirrus, a new compact car. GM is developing closer ties with Yanase and working steadily to boost sales of the German Opel, with some success. Sales of GM vehicles rose from 25,000 units in 1993 to more than 27,800 in 1994. Toyota has agreed to sell 20,000 Cavaliers a year starting in 1996. Both GM and Ford are aiming for annual sales of 200,000 units (not including the Cavalier for Toyota).

The automakers' relative success and resultant confidence in the Japanese market can be attributed largely to the following four factors:

1) The strong yen, which brings about massive reductions in import costs, is probably the major factor. In anticipation of a rise in sales volume, the automakers can reduce retail margins and cut prices and thereby sell their products in a price range comparable to that of domestic vehicles.
2) The SUV proved a major success. It had taken a great technological leap forward with the advent of the 4WD automatic transmission in the 1980s. In the late 1970s, I had the opportunity to test-drive a 4WD fitted with a prototype automatic transmission developed by one of the suppliers. I found it very easy to climb steep hills without having to change gears manually. I thought this was great and hoped that a vehicle with this transmission would come on the market as soon as possible. The advent of a 4WD automatic transmission made 4WD easy to drive even for women, and the vehicles were marketed in particular to consumers in higher income brackets. Chrysler got the best results. Ford developed the Explorer with Mazda (those sold by Mazda were named the Navajo), to be manufactured at a Ford plant in the U.S. Toyota, having made its name throughout the world with the Land Cruiser since the 1950s, was undeniably a little slow to move away from the military style toward a luxury model for city consumers. Mitsubishi was quick to concentrate its efforts in this field, and could make a solid foundation for continuing solid growth even after the bubble economy burst. These types of vehicles were invariably heavy because of their solid structure, which meant higher fuel costs for consumers. Just the other day an acquaintance was complaining about the poor fuel economy of his new U.S.-made SUV. Many Toyota 4WD owners selected diesel-engine vehicles. The real battle of SUVs between imports and domestics in Japan is now starting.
3) The advent of right-hand-drive vehicles also made a big difference. Without a doubt Chrysler's SUV really started to take off when the price was reduced and the right-hand-drive version was introduced. Japan had been arguing that U.S. automakers had made no effort to enter the Japanese market and that the lack of right-hand drive was a

typical example of their negative attitude. But some automakers countered by claiming that many Japanese prefer left-hand-drive models. The main problem with left-hand-drive vehicles in Japan is the danger when passing vehicles. You have to stay well behind the car, edge the nose of the car slightly over to the right to check for oncoming vehicles, and then accelerate very smartly to overtake. The last step requires considerable power and acceleration in order to quickly build up more speed than the car ahead. Large, powerful imported cars can do such a move successfully.

Automakers can build either left- or right-hand-drive cars without incurring significant costs if they design the car with that in mind at the initial stages. The fact that they neglected developing such a car proves that they had no will or interest in entering Japan's market. After the success of the Cherokee, Chrysler is now preparing right-hand-drive models for all vehicles. Ford has also decided to design all new vehicles since 1988 with both left- and right-hand drive. Ford's Probe, a right-hand-drive product jointly developed with Mazda, has been imported to Japan since 1993. A right-hand-drive version of Taurus is planned for the 1995 model, and for the Explorer in 1996. GM's main product, Opel, has right-hand drive. Volkswagen has been making right-hand-drive vehicles for years. The era of imported left-hand-drive vehicles in Japan is now over, and the competitive power of imports has been strengthened in Japan's market.

4) Japanese consumers are attracted to the size and power offered by imported vehicles. Until the abolition of the commodity tax and the revision of the automobile tax in 1988, most Japanese cars were small cars. The heavy tax on larger vehicles encouraged both individual and corporate owners to have cars that have no longer than 170 cm width, with engines no larger than 2000 cc. The discrepancy in tax rates was decreased considerably with the abolition of the commodity tax and the introduction of the consumption tax, while the revision of the automobile tax decreased the tax burden on owners. Consumers were able to choose the most suitable size and engine with little consideration of legislative restrictions. This led to a significant increase in the number

> of larger cars (denoted in Japan by a number plate starting with 3) being manufactured and purchased. But automakers are still firmly welded to the car sizes they have become accustomed to designing over the last few decades.
>
> As foreign cars were originally designed with little or no consideration of Japanese laws and regulations, they were very heavily taxed, even when they differed just a little in dimensions from the tax standard. But tax reform and the high appreciation of the yen have made "size-free" foreign cars attractive to Japanese consumers. The Mondeo, for instance, was taxed heavily for its size under the old system, and had little chance of enjoying widespread popularity. Although only slightly wider than the traditional small Japanese car, the Mondeo feels much more roomy and comfortable. It is now available at a price almost equal to domestic price levels and is selling well (many advance orders have already been received). This small but significant difference in size appeals to consumers.

The main issue of imports in the Japanese market from now on will be quality. On May 24 and 25, 1993, the Big Three sponsored a Drive American Quality Exhibition in Washington, where some 120 new models (40 from each manufacturer) were available for test drives by President Clinton, congressmen, and other leading government figures. (The American International Automobile Dealers Association protested the absence of Japanese cars manufactured in the U.S., but in vain.) At that time, the defect rate of Big Three vehicles within the first three months of use had dropped by 16%, as reported in the 1993 J. D. Power Initial Quality Survey (Table 12). This was taken as evidence that the quality gap with Japanese manufacturers had indeed shrunk, and that American-made vehicles were improving. The president himself took a seat behind the wheel for a test drive and noted that American vehicles were now equal to foreign vehicles in terms of quality and performance. The campaign was successful in convincing the American media that American vehicles were as good as Japanese vehicles. Indeed, the news that American cars had caught up to the Japanese in quality was spread not only in the U.S. but in Japan.

Table 12: Problems per 100 Cars, by Origin, 1988-93

	1993	1992	1991	1990	1989	1988
Big 3	113	136	148	152	159	175
Asian	94	105	112	119	120	157
European	128	158	154	158	175	237

Source: J.D. Power and Associates Initial Quality Study, *Automotive News*, May 31, 1993

Table 13: Top 1993 Models
Problems per 100 Cars

Model	1993	1992
Lexus LS 400	54	46
Lexus SC 300/400	66	72
Infiniti J30	68	NA
Infiniti Q45	70	53
Toyota Camry	70	84
Acura Legend	74	85
Ford Crown Victoria	74	NA
Lexus ES 300	75	NA
Toyota Paseo	76	NA
Buick Park Avenue	77	NA
Toyota Corolla	77	76
Toyota Tercel	77	NA

Source: J.D. Power and Associates Initial Quality Study, *Automotive News*, March 31, 1993

Table 14: Top 1993 Makes
Problems per 100 Cars

Make	1993	1992
Lexus	66	73
Toyota	74	85
Infiniti	78	70
Acura	89	115
Lincoln	89	114
Honda	92	105
Buick	95	125
Mercedes-Benz	95	BA
Nissan	95	108
Saturn	96	119
Volvo	100	BA
Mercury	101	BA
Oldsmobile	101	12
Industry Average	107	125

BA = Below Average
Source: J.D. Power and Associates Initial Quality Study, *Automotive News*

Table 15: 1993 Top Light Trucks
Problems per 100 Trucks

Make	1993	1992
Toyota	81	76
Mazda	104	119
Ford	114	BA
Nissan	120	110
Chevrolet	122	139
Plymouth	126	BA
Geo	127	119
Industry Average	129	147

BA = Below Average
Source: J.D. Power and Associates Initial Quality Study, *Automotive News*, May 31, 1993

However, a closer look at the more detailed figures released by J. D. Power (Tables 13–15) shows that American vehicles, despite some improvement, are still way behind Japanese vehicles, especially Toyota.

The April 1994 issue of the annual *Consumers Union Survey of Automobile Quality in the U.S.* showed the same conclusion. Compiled from reports received from hundreds of thousands of users throughout the U.S., the survey gives quality rankings for all vehicles by year, model, and item. The survey is the most authoritative in its field. And it is said that 80% of vehicle buyers use the report as their reference when making a purchasing decision.

Japanese vehicles maintained their superior quality, while American vehicles were not as good as they were demonstrating. In addition to a string of recalls in the U.S., all of the Big Three had suffered quality problems at production startup of their 1995 models. Once again, the Big Three have been obliged to admit that Japanese vehicles are still way ahead in terms of quality. Despite the high appreciation of the yen, which made Japanese vehicles roughly $2,000 more expensive than American vehicles of similar size and class, their sales in 1994 were even stronger than in 1993.

Japanese consumers are extremely demanding when it comes to quality. The first thing the Big Three must do if they want to maintain a longstanding and successful presence in Japan is to provide high-quality vehicles. In 1995, GM president Jack Smith reportedly urged his staff working on the 1996 Cavalier (to be supplied to Toyota) to make it equal to or better than Toyota quality standards.

The Big Three have started their move into the Japanese market with a sales target of 500,000 units a year. If this includes vehicles from other European makers, they may be able to sell 600,000 or more foreign vehicles per year in Japan. The battle for market share in Japan is a zero sum game. Any increase in imports will be at the expense of domestic vehicles. The comparative portion of Japanese marketing power is moved to imports. Not only sales outlets but also people are taken away by imports.

Because the Japanese market was isolated from foreign auto trade for such a prolonged period, dealers have developed a managing style of their own through trial and error over the last 50 years. Salespeople, for instance, are usually salaried, with minimal sales commissions. But they have considerable potential for rising to a supervisory or managerial position, and if they work hard, lifetime employment is guaranteed. Japanese dealer margins, including sales incentives, are about 10% higher than in the U.S., reflecting the high level of investments in land, buildings, equipment, and education. A considerable portion of salespeople's work hours are consumed preparing documents for government agencies. Therefore, sales efficiency or productivity is inherently low in Japan. Furthermore, fierce sales competition forces them to take costly measures such as price cuts (in many cases, inflating trade-in value).

One wonders what changes we will see in our automobile dealers when Americans bring in their own operating style, such as wages based on sales results or employment based on contracts rather than lifetime employment. In order to take root and develop in the Japanese market, the most important task for foreign automakers is to demonstrate that their sales outlets can turn out profits operating in either way, Japanese or American.

In 1994, domestic cars and imports competed freely for the first time in the Japanese market. The arrival of the black ships has indeed shaken automobile dealers.

The Automobile Industry and the Bubble Economy

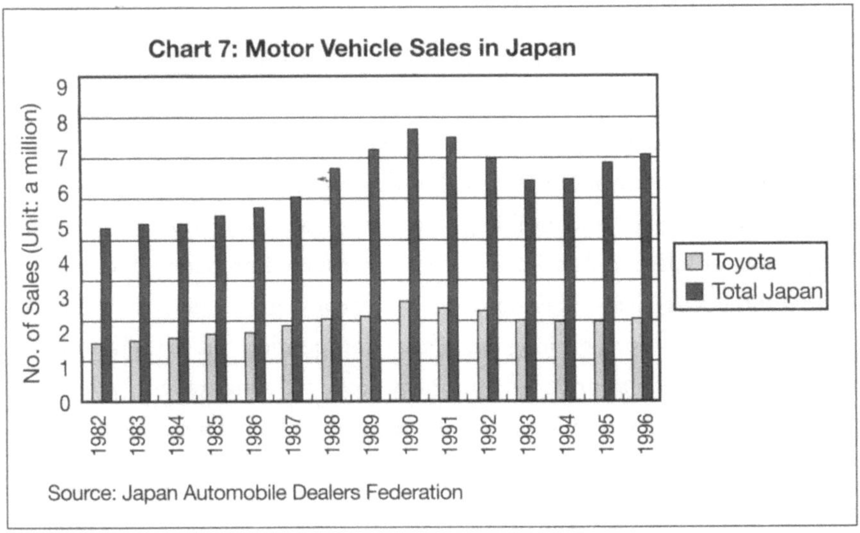

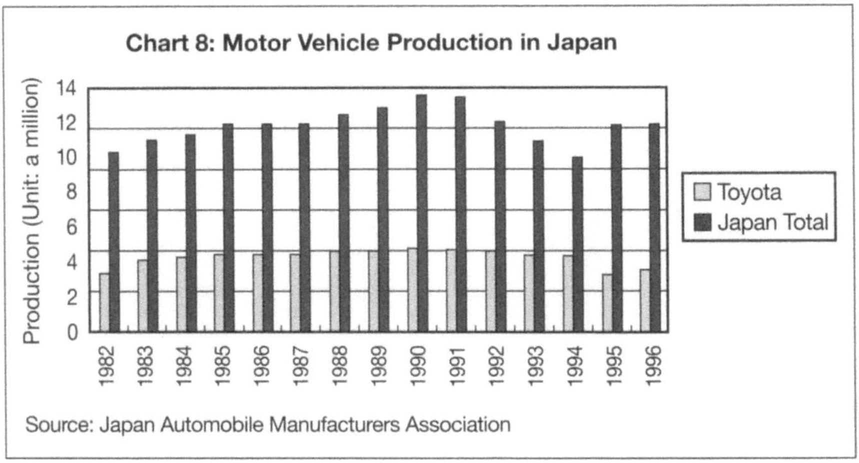

During the period from 1987 to 1991, Japan's car market also underwent abnormally high growth along with the swelling of the bubble economy. The onset of the bubble economy in 1987 triggered an explosion in the domestic auto market, which had been expanding steadily at a stable rate of 2% or 3% per year.

In 1990, the auto market consisted of over 7.7 million units, as compared to 5.7 million in 1986 (directly before the bubble), an average yearly increase of 500,000 units. When the bubble economy burst in the early part of 1991, the market suddenly took a downward dive. It dropped to 6.47 million units in 1993. There followed a slight recovery to 6.53 million units in 1994, but this was due partly to some 300,000 units of imports (Chart 7).

Production output rose from 12.2 million units in 1986 to 13.5 million units in 1990, an increase of more than 1.2 million units over the four-year period. The subsequent fall was severe indeed; in 1994, output had dropped to barely 10 million units, the same level as more than a decade before. The main causes were declining domestic sales, a slump in exports due to the strong yen, falling domestic production levels because of greater offshore production (referred to as "hollowing out"), and rising imports. Annual production at Toyota reached 4.21 million units in 1990, an increase of 550,000 units since 1986. As Chart 8 shows, Toyota was affected like everyone else when the bubble economy collapsed, with production falling to 3.51 million units in 1994, the level of ten or more years earlier.

Note: According to an Economic Planning Agency release dated November 17, 1994, the bubble economy lasted 53 months, from December 1986 through April 1991, while the "Heisei Recession" is defined as the 30-month period from May 1991 through October 1993.

Table 16: Investment for Plants and Facilities by Japanese Automobile Industry (Unit: Billion Yen)

	1986	1987	1988	1989	1990	1991	1992	1993	1994	1995
Motor Vehicle Manufacturing Industry	1,246	1,031	1,278	1,701	2,300	2,184	1,970	963	771	786
Automaker	900	734	924	1,183	1,575	1,593	1,492	714	636	636
Body Building	42	50	72	97	133	91	72	55	44	55
Parts	304	247	282	421	592	500	406	194	92	95

Source: Survey of Investment for Plants & Facilities by Leading Industries by MITI

During the bubble economy period, manufacturers boosted investment in plants and equipment (including overseas investments) to nearly twice the normal levels (Table 16). The funds were obtained cheaply through issues of warrant bonds and convertible bonds, riding the wave of the economic upturn (Table 17). When repayments on warrant bonds fell due, right in the middle of the collapse of the bubble economy, manufacturers were forced to quickly generate more funds by issuing corporate bonds, selling off fixed assets, and realizing short-term securities such as shares held as designated cash trusts. At the same time, they had to considerably scale down current investments in plants and equipment. All the automobile manufacturers were hit hard by the Heisei post-bubble recession, when appreciation of the yen was accompanied by a downturn in sales, and revenue plummeted sharply.

Table 17: Convertible and Warrant Bonds Issued by Automakers (Unit: Billion Yen)

	Convertible Bonds	Warrant Bonds	Total
Nissan	160	270	430
Honda	90	90	180
Mitsubishi		50	50
Toyota	500	330	830

Source: Security Reports
Warrant Bonds were issued by Nissan in 1986 and 1989, by Honda in 1989, by Mitsubishi in 1990, by Toyota in 1987 and 1989. Convertible Bonds were issued by Nissan in 1981, 1983 and 1987, by Honda in 1982, 1983 and in 1985, by Toyota in 1996 and 1988.

A closer look at the bubble period shows that spectacular growth in sales was not reflected in profits, illustrating the intensity of the struggle among manufacturers competing for market share. When the market was expanding rapidly, smaller manufacturers saw their chance to catch up to the bigger ones, and began to boost production and sales capacity by modernizing equipment and expanding their sales networks. Meanwhile, the larger manufacturers, keen to preserve market share, put pressure on their dealers to sell, with the result that sweeping price reductions became common at retail outlets. So although the sales volume was rising, the automakers had to compensate financially stricken dealers for their losses by providing a considerable amount of sales incentives, and as a result they could not enjoy the benefits of the bubble economy before the recession came. The bubble economy left a deep scar on the automobile industry and only served to widen the gaps among manufacturers.

Only Mitsubishi, successful at positioning the SUV as its main product, was able to sustain reasonable profits throughout the bubble economy period and even in the postbubble recession could maintain the healthy growth in sales. Honda steadily built up a foundation in North America but did not seem to be aggressive enough against its competitors at home. The new Accord, announced in 1993, was well received both at home and abroad, particularly in North America, where once again it vied with the Ford Taurus for the title of top mass-produced vehicle. Mazda was suffering the most and now appears to be more dependent on Ford, which owns a 25% stake. Nissan, which had been gradually losing market share under pressure from other manufacturers, was forced into even greater difficulties during the bubble economy period. Sales levels subsequently made a comeback (despite the strong yen) with the release of the Altima (Bluebird in Japan), a very popular model in the U.S.

After posting ordinary profits of 733.8 billion yen in 1990, Toyota suffered a massive setback. To some extent this can be attributed to the strong yen and other factors common to all manufacturers, as previously described. In addition, though, the new models Toyota released at that time did not perform as well as expected. A good example was the new Crown, introduced in 1990. Toyota added a luxury model, the Majesta, to the Crown series and set up a highly automated line for the production of new Crown models. But their basic style did not appeal to business users. Furthermore, Toyota failed to develop grades among the series and suffered the ignominy of slow orders from the outset.

Similarly, the T100 pickup truck, developed especially for the U.S. market, sold poorly despite good reviews, creating a huge ruckus in the American print media about Toyota's alleged "failure." Farmers in the Midwest, at whom the vehicle was targeted, remained firmly attached to their cheap, antiquated, but powerful GM and Ford pickups. Clearly the market did not need a stylish but expensive alternative.

Prior to this, the late Tsutomu Oshima, then Toyota's vice chairman, and vice president Tatsuro Toyoda had come to realize that the Product Planning Office was no longer functioning effectively after the merger of Toyota Motor Sales and Toyota Motor Company. In 1990, the Product Planning Office was transformed into the Product Planning Division under the direct supervision of the president, to give it a greater role. The RAV4, a small SUV, which met with acclaim when released in 1994, and the new Vista and Camry models after that, were developed in accordance with a clear marketing strategy under the new organization.

During the bubble economy period, Toyota strengthened offshore production facilities as a bulwark against currency appreciation. In the U.S., a new engine plant was built at TMMK, and its body assembly plant was enlarged. An engine plant and a body plant were constructed in the United Kingdom, too. Local parts procurement was stepped up for Toyota plants in the U.S., and a greater effort was made to use American-made parts in Toyota factories back in Japan.

Mitsubishi left the competitors far behind through the bubble period. Honda and Toyota managed to see the light when the post-bubble recession ended. Mazda had a hit product, the Familia, and Nissan the Cefiro. Both were expected to recover in 1995. In the immediate future, as the Japanese market is unlikely to expand the way it did during the bubble period, the vacuum created by the failure of any manufacturer could well be filled by imported vehicles. For this reason alone, domestic automakers must overcome the after-effects of the bubble economy as early as possible.

The Japanese economy of today looks better than it really is, supported by the booming economy in the U.S. Sometime in the future their economy will experience a downward trend. This is the decisive point for the Japanese economy. U.S. manufacturers have been slashing fixed expenses and creating lean management structures that will enable them to survive even when the next economic recession comes. In Japan, meanwhile, the management attitude generally is to hang on the current economy and wait for things to improve. Japanese and American automakers alike must prepare for the next economic downturn by becoming leaner and stronger.

Developing the Technology Basis

The prosperity of the U.S. in the 1950s was supported by people of the middle class. In the 1970s, however, people in the middle-income bracket began to decrease, splitting into a small group of high-income earners and a much larger population of low-income earners. The main reason for this was said to be the decline of the manufacturing industries. Similarly, Japan's current prosperity was brought about by the growth of manufacturing industries like the automobile industry and also by the resultant increase of people in the middle-income brackets. If Japan is to stay prosperous in the 21st century, we must ensure the continued development of the manufacturing industries.

The basis of industry is endless technological development, promoted by engineers. Japan's automobile industry has cultivated the finest world-class engineers. The engineer groups of Toyota, Nissan, Honda, Mitsubishi, Mazda,

Isuzu, and others are the envy of the world's automakers. Backed up by highly trained, highly skilled factory workers, these engineers have created top-quality cars and highly efficient production systems that enable Japan to maintain strong sales in the U.S. despite the strong yen.

Although GM chairman Smith is on the right track, the faltering progress of his efforts to rebuild is clearly due to the company's short supply of in-house engineers. Ford's decision to merge the European and U.S. technology development divisions was an attempt to compensate for the dearth of talent in North America. But this has not been enough. Ford is now considering how best to utilize the technical expertise at Mazda.

As I have tried to explain in this book, Japan's engineering predominance has brought prosperity not only to its automobile industry but to all Japanese manufacturing industries. Yet today we see an alarming phenomenon among students of elementary and middle schools, who will support the technological base of the next generation. They are turning away from science and mathematics. Those who like these subjects tend to be treated as outcasts. More and more universities are choosing to omit mathematics from their entrance examinations in order to attract more students, which only exacerbates the trend. Students of middle and high schools are apparently convinced that the best way to enter the university of their choice is to abandon mathematics and science and concentrate on English and the social sciences. I am afraid that we are now facing a possible decline in academic proficiency among those students.

In July 1994, the Japan Association of Mathematics and three other academic associations issued a joint statement entitled "The Crisis in Mathematics Education," which blames the Japanese cramming-style education for the fact that students are drifting away from mathematics and science, and also for the decline in mathematical ability. Shigeru Iidaka, chairman of the Japan Association of Mathematics, warned, "The tendency to underrate mathematics will bring a level-down of logical thinking power necessary not only for the study of natural science but also for social science." This was preceded by a similar statement from the Japan Association of Physics and two other academic associations in April, calling for more attention to science education (*Nihon Keizai Shimbun*, July 3, 1994). All institutions concerned with education in Japan, from universities down to elementary schools, will, together with politicians and industry, take decisive measures to cope with the crisis as quickly as possible.

According to the July 23, 1994 issue of *The Economist*, computer manufacturers are far more powerful than software makers, which have made spectacular progress. Comparing the software giant Microsoft with Hewlett Packard, one of the leading manufacturers of computer and associated equipment, *The Economist* found that the latter was much bigger in terms of overall sales. This is proof that the manufacturing industry is sustaining the prosperity of a nation.

Although Japan's present prosperity has been brought about by the export industries, those which are competitive in the world market are limited to a few manufacturers such as electronics and automobiles. The demand for vehicles in the 21st century will grow quickly in China, India, and other Asian developing countries, where motorization is expected to expand rapidly. Technological innovations are still under way in both the electronics and automobile industries. In order to meet increasing demand, the world automobile industry is now competing to supply vehicles that are safer, more comfortable to drive, and less harmful to the environment, at the lowest possible cost, by making good use of the most advanced electronics technology. Japan must fortify the technology base supporting its manufacturing industries if Japan is to avoid the type of decline seen in the U.S. in the 1980s and 1990s, which led to rising unemployment, declining income levels, and a growing national deficit.

People of my generation believed without doubt that "the later you are born, the better you enjoy life." This has been proven true. I hope people coming after us have the same belief and that it will be proven true.

A d d e n d u m

Addendum

Changes in the world of the auto industry occurred so quickly over the past two years since the publication of the first Japanese edition of *The Origin of Competitive Strength*, I found it necessary to add this supplementary chapter to the English version.

Four major issues are addressed here. First is the ongoing controversy with regard to auto trade between Japan and the U.S. Second is the faster-than-expected development of such "zero-emission" vehicles as battery-powered electric vehicles. A third issue is the growing competition among mini-vehicle manufacturers. And the fourth issue is the remarkable development of electronic science and its application to the industry.

The world auto industry is preparing itself to meet the demands of the 21st century, when many more vehicles will be required to sustain the fast-growing global population. The industry and its products must become more friendly to people and the environment by reducing fossil fuel consumption, minimizing emission, protecting people from traffic accidents and by supplying more affordable vehicles. While the existing auto companies are making every effort to survive through the next century, new manufacturers will be born and grow up in regions where none exist today. It is quite probable that continued vigorous competition among auto manufacturers will bring better life to people through safer, cleaner, and less expensive vehicles.

Japan-U.S. Auto Trade Negotiations

On May 9, 1996, Toyota Motor Corporation president Hiroshi Okuda (who succeeded Tatsuro Toyoda as president in August 1995, after the latter suffered a stroke) announced his company's plan to build a new $400-million engine plant in Putnam County, West Virginia. In the announcement Okuda said,

> "The new plant will build engines for use in North-American-built Corollas in late 1998. The plant will have an annual production capacity of 300,000 units. Once it reaches full production, we expect that it will bring 300 new jobs to the Mountaineer State..."
>
> "Today's announcement of this new engine plant is yet another step in the localization of all aspects of the vehicle production process.

By 1998 we plan to have the capacity to produce 1.2 million vehicles here in North America..."

On the previous day, president Okuda attended the groundbreaking ceremony for Toyota's fourth North American assembly plant in Gibson County, Indiana. He indicated that the $700-million plant, to be completed in the fall of 1998, would produce approximately 100,000 full-sized T100 pick-up trucks per year. The plant will bring Toyota's North American production capacity to over 1.1 million vehicles per year by 1998 and will employ about 1,300 workers, raising the total number of Toyota employees in the U.S. to more than 20,000.

These actions provide evidence of Toyota's commitment to executing the new global business plan the company announced in June 1995, which outlined sweeping new investments in both the U.S. and across the globe.

The announcement of voluntary business plans by Toyota and four other Japanese automakers coincided with the settlement of an auto trade dispute which threatened to expand into a full-fledged auto trade war between Japan and the U.S.

The U.S. trade negotiators had demanded numerical targets from the Japanese government for procurement of U.S.-made auto components and parts by Japanese automakers for use not only in the U.S. but in Japan. The Japanese government insisted that the parts trade between automakers and suppliers was purely a business-to-business matter, and that government had no role to play. Furthermore, the government believed that setting numerical targets for trade between the governments was against the trade rules of the newly established World Trade Organization (WTO). In spite of repeated negotiations, the two countries failed to reach consensus. Then, on May 16, the U.S. announced it would impose a 100% tariff on 13 Japanese luxury car models, including Lexus, based on Section 301 of the Trade Act of 1974, and set a deadline for negotiations — June 28. On May 18, Japan filed suit with the WTO and began to prepare retaliatory measures against the U.S. sanctions.

Meanwhile, U.S. Ambassador to Japan Walter Mondale, U.S. trade representative Mickey Kantor, Minister of International Trade and Industry Ryutaro Hashimoto (Japan's present Prime Minister), Toyota Motor Corporation chairman Shoichiro Toyoda, the late U.S. Commerce Secretary Ron

Brown and related officials of both governments made efforts to reach an agreement. On June 28, the deadline set by the U.S., the two sides found a way to finalize an agreement with neither government losing face.

Among the voluntary business plans announced by Japan's five major auto manufacturers, Toyota's was the most wide-reaching and aggressive, and U.S. officials paid close attention to it. The business plan contained a key clause: "Through the following activities, Toyota will work to continually receive the tariff exemptions set forth by NAFTA." In order to qualify for those exemptions, automakers in North America must localize the procurement of components and parts up to 50% by 1997, 56% after 1998, and 62.5% by 2002. U.S. officials believed they were able to calculate an estimate of Japanese procurements of U.S. components and parts using these localization rates, even though neither the Japanese automakers nor the Japanese government had announced exact procurement figures. Thus Japan could maintain its position that it was against setting numerical targets, while the U.S. could announce its own estimated figures. On August 16, 1995, Japan and the U.S. finalized the auto trade agreement they had reached in Geneva on June 28.

An *Automotive News* editorial on October 7, 1995, summed up the outcome: "The trade war has been averted, and there will be peace in our time, or so the politicians say. It is good that no blood was spilled. At the deadline, Japan gave a little, the United States gave a little..."

Toyota's recent activities in the U.S. bear witness to its intention to follow its global business plan. On September 18 and 19, 1996, Japan and U.S. governments held a meeting in San Francisco to review the implementation of the Japan-U.S. Auto Trade Agreement. The meeting made it clear that both governments were satisfied with progress of the implementation.

The Office of the U.S. Trade Representative submitted a report to President Clinton on October 21, 1996, saying that implementation has resulted in significant progress toward achieving the agreement's key objectives of improving access to the Japanese vehicle distribution system, eliminating regulations that restricted access to the parts aftermarket in Japan, and improving opportunities for U.S. original equipment parts suppliers in Japan and with Japanese transplants in the U.S.

The USTR report illustrated the significant progress with the following examples:

- Sales of North American-produced Big Three motor vehicles in Japan have increased by 40% during the first half of 1996 compared to the same period in 1995. The growth rate of 1996 exceeds that recorded for imports from Europe (11%) and vehicle imports overall (15%). This growth is particularly noteworthy, given that Japan's economy and vehicle market have been stagnant.
- Exports of U.S. automotive parts to Japan have increased 60% since 1992, growing from $1 billion in 1992 to $1.6 billion in 1995. During the first six months of 1996, exports totaled $901 million, an 11% increase versus the same period in 1995.
- Numerous U.S. parts companies that were previously unsuccessful in Japan now report significant new contracts and sales opportunities.
- Over the past year, the Big Three have established 114 new sales outlets in Japan — an increase of 84 outlets since April. The addition of new sales outlets strengthens marketing capabilities for Chrysler, Ford and General Motors in Japan.

Progress has been made, but there is more to be done, and the U.S. government shares the industry's concerns about the slow pace of improving access to the vehicle distribution system in Japan, according to the report.

U.S.-Japan Motor Vehicle and Automotive Parts Trade

	1992	1993	1994	1995	1996
Motor Vehicles (Units)					
Exports	51,329	68,754	118,923	166,012	161,549
Imports	1,875,459	1,752,065	1,764,246	1,478,117	1,242,533
Motor Vehicles (in millions of dollars)					
Exports	887.717	1,157.127	2,030.340	3,107.727	2,673.856
Imports	22,341.732	23,336.960	25,977.134	23,019.544	20,900.246
Balance	−21,454.015	−22,179.833	−23,946.794	−19,911.817	−18,226.392
Automotive Parts (in million of dollars)					
Exports	1,035	1,130	1,485	1,637	1,965
Imports	10,817	12,339	14,335	14,667	13,409
Balance	−9,781	−11,210	−12,849	−13,030	−11,444

Source: U.S. Department of Commerce, Office of Automotive Affairs, Motor Vehicle Division

The preceding table clearly shows that U.S.-Japan auto trade is improving. The U.S. import of motor vehicles from Japan decreased dramatically since 1994, while U.S. vehicle exports to Japan increased gradually. These results were due mainly to localization of Japanese vehicle production. Furthermore, Japan's manufacturers replaced the imports with the local products and dared not to do parallel supply of imports and local products. This also contributed to the above results. Parts imports from Japan went down because Japanese manufacturers increasingly procured parts locally. All this progress was made business-to-business — not by governments.

Dr. Shoichiro Toyoda, chairman of Toyota, who also serves as chairman of the Keidanren (Japan Federation of Economic Organizations), gave a speech on October 16, 1995 at a dinner meeting of the Japan-America Society of Washington, D.C. Discussing the recent Japan-U.S. trade negotiations and the pressure applied on the U.S. government by American corporations, Dr. Toyoda referred to a sentence in Henry Ford's autobiography: "'Our help does not come from Washington, but from ourselves,'" he said, "I think business executives in Japan and the U.S. should make the effort to build mutually favorable relationships by concentrating on their corporate activities rather than unnecessarily depending on their governments."

Meanwhile, both the U.S. Big Three and Japan's auto manufacturers are expecting big profits for 1996 fiscal year.

Zero-Emission Vehicles

In 1900, 4,000 motor vehicles were produced in the U.S.; 40 of them were powered by steam engines, 38 were electric and the remaining 22 were gasoline-powered cars. Since then, vehicles driven by gasoline-powered internal combustion engines grew to dominate the worldwide motor vehicle market because the electric vehicles powered by lead-acid batteries had a wall that existing technology could not breach. The driving range per charge of electricity was simply not enough to make such vehicles practical. Furthermore, due to the steady supply and relatively low cost of oil, electric vehicles lost their market to vehicles powered by internal combustion engines.

In 1925, Sakichi Toyoda, the inventor of the Toyoda automatic loom and founder of Toyoda Automatic Loom Works Ltd., made an agreement with the Imperial Invention Association specifying the contribution of 1 million yen (according to an estimation, about $15 million in today's dollar value). Toyoda believed that in order to cope with the lack of oil resources, Japan had to expedite the development of alternative fuels for vehicles. Toyoda hoped his contribution would promote the research and development of electric batteries. Kiichiro Toyoda, the founder of Toyota Motor, followed in his father Sakichi's footsteps and continued the research and development of batteries. In 1935, he built six experimental electric cars. Their driving range was 60 km (37 miles) per charge. Kiichiro Toyoda used one of them himself for commuting to his laboratory from his home.

But in 1938, the government asked auto manufacturers to halt production of passenger cars including electric cars in order to save materials for military use and no one produced them again for many years.

After the surrender of Japan to the Allied Forces in 1945, it became the urgent necessity for Japan to find ways and means to offset the lack of oil supply from overseas. Manufacturers made efforts to develop vehicles powered by alternative fuels such as charcoal or carbide. The electric vehicle was given top priority among them. At that time, I was working for the Ministry of Commerce and Industry (present Ministry of International Trade and Industry) in its Automobile Section. One of my duties was to promote the production of electric vehicles. It was a time of a scarcity of materials. The supply of rubber tires, for example, was so limited that without getting government allocation of tires, auto manufacturers could not produce vehicles at all. Because of this, allocation of rubber tires became the best incentive to encourage manufacturers to produce electric vehicles. A further incentive was that while production of passenger cars was prohibited by order of the Allied Occupation Forces, electric cars were not counted as passenger cars and thus were exempt from the manufacturing ban.

Standardization of electric battery charging stations, and plugs and sockets of power lines was encouraged. The Ministry put the Society of Automobile Engineers of Japan in charge of testing electric vehicle performance in order to accelerate the improvement of vehicle quality. Until 1949, the production of electric cars in Japan was more than that of conventional cars powered by conventional engines.

In 1949, the Allied Occupation Forces approved the restart of passenger car production without conditions. In the next year, 1950, the Korean War broke out and immediately after that the global price of lead rose suddenly to a level that effectively drove the cost of lead acid batteries (and the vehicles powered by them) out of the range of affordability. Manufacturers stopped the production of electric vehicles and they vanished from the scene.

At the end of 1949, Toyota separated its electrical components division from the company and established a new and independent company named Nippon Denso (presently DENSO). At the first stockholder meeting of DENSO, the stockholders voted to add "production and sales of electric vehicles, electric equipment" to its certificate of incorporation. They started to sell electric cars in July, 1950. The DENSO vehicle's driving range per charge of electricity was 195 km (121 miles). In the month when the car was introduced, the Korean war broke out. DENSO was obliged to give up the production of electric vehicles because of the sudden rise in lead prices. The company built and sold 50 units, then stopped the production.

In the 1970s, Japan suffered greatly from the worldwide oil shortages. The 1973 Arab oil embargo and the 1979 price shock caused by the Iranian revolution hit Japan hard. With no domestic oil supply, Japan's national economy was damaged so badly by the oil shocks that it took a few years to recover. The Ministry of International Trade and Industry resumed the promotion of research and development of electric vehicles and batteries as one of the government policies designed to reduce the effects of oil shocks and established an industry-government consortium in the early 1980s. Toyota, Daihatsu and Hino worked closely with the consortium and other manufacturers, building a few experimental vehicles.

Thus the research and development of electric vehicles and batteries has become a traditional area of interest and expertise for Toyota and Toyoda family.

Perhaps the most tangible result of Toyota's electric vehicle efforts has been the mass-production of battery-powered forklifts. These zero-emission lifts are very popular for use inside factory shops and warehouses. Today more than 60% of Toyota forklifts are electric, a real success in the commercial production of electric vehicles.

In 1990, the State of California passed a so-called "zero-emission" law requiring principal car makers to sell a certain percentage of zero-emission vehicles beginning in the 1998 model year. (In the spring of 1996, the law was reviewed and revised to delay the required introduction of zero-emission vehicles to the 2003 year model.) The law stimulated the principal car makers — both the U.S. Big Three and foreign nameplates — to make their utmost effort to develop zero-emission vehicles.

GM had also been working on research and development of electric vehicles. As soon as the California law was passed, GM started to build an experimental electric two-passenger vehicle named Impact. The automaker invested $350 million to develop the Impact. It took more than six years before the Impact went on sale in November, 1996 in Southern California and Arizona. Although expected sales volume has not been announced, observers expect GM will sell just a few hundred units a year.

The GM electric vehicle, sold as the "EV-1," boasts a driving range per electricity charge of 70 miles city, 90 miles highway, with lead-acid batteries. Its top speed is 80 mph. It accelerates from zero to 60 mph in 8.5 seconds. The car requires between 12 and 14 hours to recharge from a standard 110-volt outlet.

The car seems to have enough performance for urban driving. But one apparent problem is its price of $35,000. GM is currently offering only a lease on the car.

In September 1996, Toyota started to sell a new RAV4 four-passenger vehicle in Japan powered by nickel-metal-hydrate batteries. The new vehicle is to be sold in California starting in the fall of 1997. Its weight is 3,212 lb., about 700 lb. more than the conventional model. The maximum speed is 78 mph and its driving range per charge is 134 miles. Its price is 4,950,000 yen (about $45,000), a little more than three times the price of a base conventional RAV4. Toyota aims to sell 100 units a year.

The RAV4's nickel-metal-hydrate batteries can be charged from ordinary household outlets by the car's built-in charger, or from recharging stands. The battery pack is located under the vehicle's center of gravity, leaving enough interior room for four adults.

Toyota has teamed up with Matsushita Electric Industries Co. and Matsushita Electric Batteries Co. to develop, produce, and market nickel-metal-hydrate batteries. On December 11, 1996, the companies announced the establishment of "Panasonic EV Energy Co." to supply batteries. The company's production target is said to be from 5,000 units to 10,000 units during the period of 1997 through 1999. Other Japanese auto manufacturers are also entering the electric vehicle market. Beginning in 1997 Honda plans to sell 300 units of its new electric vehicle powered by nickel-metal-hydrate batteries over three years. Nissan plans to sell electric vehicles on a limited basis next spring in Japan. In a joint venture with Sony, Nissan has also been making efforts to introduce electric vehicles powered by lithium-ion batteries. The Nissan/Sony lithium-ion battery was given the "Best of What's New" award by *Popular Science* in 1996. Hedging its bets on the lithium-ion battery, however, Nissan made arrangements in December 1996 to purchase nickel-metal-hydrate batteries from Toyota and Matsushita.

On October 4, 1996, Toyota announced that it had succeeded in developing a fuel cell electric vehicle (FCEV) based on the RAV4. The vehicle is powered by a fuel cell generating 20 kilowatts of electricity through a chemical reaction between hydrogen and oxygen. The fuel cell's only by-product is water. The FCEV boasts a driving range of 155 miles per charge of hydrogen and a maximum speed of 62 miles per hour. Toyota built two experimental fuel-cell-powered RAV4s, costing 100 million yen (about $910,000) each. In May, 1996, Mercedes-Benz announced the world's first fuel-cell-powered vehicle. But while the Mercedes carries two hydrogen gas cylinders on its roof, Toyota's new FCEV stores both the fuel cell and hydrogen storage inside a standard RAV4. This innovation is due to a new hydrogen absorbing alloy developed by Toyota that has doubled the hydrogen storage capacity of existing alloys. Toyota researchers hasten to point out, however, that the FCEV is only in the experimental stage and it would take years of further development before this vehicle is put to practical use on the road.

The 13th international electric vehicle symposium (EVS-13) was held in Osaka, Japan from October 14 through 16, 1996. All the Japanese makers, the American Big Three and France's PSA/Peugeot-Citroen were represented. But most German, Swedish and Italian car makers did not attend.

Although this symposium has been held every two years in different countries, EVS-13 was a historical event for the world auto industry. All delegates left with a belief that after more than a century, the world auto industry has at last succeeded in bringing electric vehicles into the mainstream. The symposium made it clear that the nickel-metal-hydrate battery was the most probable one for widespread near-term use. While the lithium-ion battery is also very attractive compared with others, safety and raw material supply issues make its broad adoption by the industry highly unlikely.

Dr. Shoichiro Toyoda, chairman of Toyota Motor Corporation, stressed in the conference's keynote speech the importance of electric vehicles as we approach the 21st century. Dr. Toyoda also noted that while further technical development will play an important role in the future of electric vehicles, cost performance is no less important. Today's electric vehicles cost more than three times that of conventional vehicles, he said, but if manufacturers could supply them at one and a half times the cost of gasoline-powered vehicles, sales would take off. Dr. Toyoda stressed that he believed this cost goal is attainable.

According to the October 21, 1996 edition of *Automotive News*, Robert Stempel, chairman of Energy Conversion Devices Inc. and former chairman of General Motors, attended the symposium and predicted the revolution may be upon us. "I believe the U.S. public will have a wide variety of EVs to choose from in the year 2000, from U.S., European, Japanese, Korean and Chinese manufacturers."

The Progress of Electronic Science's Application to the Auto Industry

To understand the progress of electronic science's application to the auto industry these past two or three years, let us look at two examples — CALS (Computer-Aided Logistics Support) and ITC (Intelligent Traffic Control).

Computer-Aided Logistic Support (CALS)

When Boeing started to build its newest 777 jet in 1986, it introduced two linked computer systems: CATIA (Computer-graphic Aided Three-dimensional Interactive Application), a Dassault/IBM program, and EPIC (Electronic Preassembly in the Computer), the add-on program devised by Boeing that allows different components to be designed and integrated into one vast computer simulation of the whole plane. Thus the company could "build" a plane without making mock-ups, based entirely on 3-D computerized data (Karl Sabbagh, *Twenty-First-Century Jet: The Making and Marketing of the Boeing 777*, Scribner, 1996).

Since then, many world automakers have tried to adopt computerized design processes similar to those used by Boeing, allowing some to develop new vehicles in less time, with less money, and requiring fewer prototypes. Development time for new vehicles, which once took more than six years in the U.S. and more than four years in Japan, has been remarkably shortened. Some auto manufacturers have boasted that they could introduce new vehicles to the market less than 18 months from the start of development. Today, due in part to such computer technologies, there is no substantial gap between Japanese and U.S. manufacturers in new vehicle development time.

Meanwhile, a new problem has arisen. Suppliers with multiple customers are forced to handle a variety of systems. Having standard computerized development systems would benefit not only suppliers but also the auto manufacturers themselves. For example, as the world metal die-making capacity is limited, automakers have to procure metal dies from manufacturers who have supply capacity at the time the dies are needed. A standardized computer system would make meeting the demands of different manufacturers with different product requirements faster and more efficient.

In order to deal with these problems, Japanese automakers—Toyota, Nissan, Honda, Mitsubishi, Mazda, Suzuki, Yamaha (a motorcycle manufacturer and auto engine builder), and Kawasaki (a motorcycle manufacturer) organized a "V-CALS consortium" in May 1996 (V stands for vehicle). Its object is to establish standards of electronic data exchange for commerce between automakers and suppliers. In cooperation with auto parts and components manufacturers, machine and facility manufacturers and electronic computer manufacturers, the consortium's goal is to promote electronic dealings on its network. Auto suppliers such as Denso, Calsonic and Ichikou Kougyo, Takada Kougyo, and Zexel became members of the consortium, and computer makers such as Fujitsu, IBM Japan, Japan Unisys, and Hitachi joined to assist in systems development. The consortium will act as a "virtual" corporation on the network. Heiichi Hamaoka, managing director of Nissan Motor, was elected president of the virtual corporation.

If V-CALS was to spread throughout the world auto industry, the industry in the 21st century would be somewhat different. It would be easier, faster, and less costly to develop and produce new vehicles. Automakers would be able to supply their products to consumers at an affordable price and at the very time they were needed. But it is feared that the flood of new products might cause brutal competition among automakers and their suppliers. We must acknowledge that developing vehicles by computerized methods will not necessarily mean success in the market. Neither will it eliminate start-up problems. V-CALS is not a panacea, but merely a tool for developing new vehicles. Such a system is absolutely necessary, however, for the automakers and their related manufacturers to survive.

Intelligent Transportation System (ITS)

The *Nikkei News Service* on May 5, 1996, featured a short article titled "Car Makers to Join U.S. 'Automatic Driving' Project." It stated:

> "The U.S. government will begin a 10-year project this autumn to develop a futuristic infrastructure for the driving of automobiles. Private corporations from not only the U.S. but also Japan, such as Toyota and Nissan, may be invited to join in the study. The project aims to integrate state-of-the-art communications and electronics technology to develop equipment and a system for guiding and controlling traffic. Led by the Department of Transportation, it will involve the Pentagon, the Department of Commerce, and the National Aeronautics and Space Administration. Detailed plans will be worked out based on proposals from Rockwell International Corp. and Lockheed Martin Corp., two U.S. companies. [The proposals are expected to be announced at the world conference on ITC in the fall of 1996.] According to U.S. authorities, more than 100 companies are likely to take part. Atlanta or Houston is projected to be the first to test the new system, the sources added. Because of the accelerated pace of increase in the number of automobiles in the U.S., driver-dependent driving will soon reach its max. The new system should enhance safety and transport efficiency, the sources added."

Although it would take years to put "automatic driving" to practical use on highways, the world of intelligent transportation can give commuters information on weather, traffic jams, and construction projects. Vehicle navigation systems that guide motorists to their destination have become very popular in Japan during the past few years not only among luxury-car buyers but among young drivers. Although the systems cost more than $3,000 three years ago, sales have been good. Manufacturers not only of electronic equipment but also of audio equipment have begun production. Competition in this new market has become so fierce it has driven down the price of a navigation system down to less than $1,000. Production has been expanded to about 550,000 units in 1995, a 60% increase over the previous year.

The VICS (vehicle information and communication system), which supplies drivers with road and traffic information through on-board car navigators, was introduced in April 1996, around the Tokyo metropolitan area.

As for so-called intelligent highway systems, the Machinery Laboratory of MITI began a study of automated car driving in the late 1960s. The Ministry of Construction will make trial runs of cars with automated driving systems in the spring of 1997 on a newly constructed highway before it opens for public use. Thus Japan would be able to contribute substantially to the ITS projects proposed by the U.S.

V-CALS as well as ITS need the cooperation of world automakers, their related manufacturers, and the governments concerned. International standards are the ideal target not only for manufacturers but for car users.

Increased Competition of Mini-Cars

Volkswagen president Ferdinand Piech announced at the Vienna International Motor Symposium in May 1996, that VW's forthcoming mini-car had been designed to be assembled in seven hours. Because of the simple assembly plan, the project will be profitable (*Automotive News*, May 6, 1996).

Micro Compact Car AG., a joint venture of Mercedes Benz and Swiss watchmaker Swatch is scheduled to put on the market a brand-new mini-car in 1998. It is said that this car will be put together from a dozen key modules and assembled in less than five hours per car (*Forbes*, May 6, 1996).

These assembly-hour cut plans are made possible by adopting modular production principles, a theory advocated for years by the controversial automotive visionary Dr. J. Ignacio Lopez de Arriotura. Lopez was recruited by Piech to VW from GM in 1993, and resigned from VW in 1996.

In the 1970s and 1980s, adoption of so-called "lean production" methods — pioneered by Toyota — increased auto production efficiency remarkably. Lean production is mainly based on "design-in" and "Just-In-Time" management. The former principle involves selecting suppliers who have the engineering as well as financial capacity to work closely with the factory to design parts and components from the initial stage of new vehicle development. The latter principle respects the ability of workers and saves wastes of time, money, materials and labor as much as possible.

The new modular production principles, advocated by Lopez, aim to increase production efficiency by perfecting lean production. The number of suppliers is limited to 10 or fewer. Suppliers assemble parts under the same roof as the factory's vehicle assembly line.

Assume, for example, that Bosch supplies starters to Volkswagen. In a modular production factory Bosch workers would assemble starters on the VW factory floor alongside the line assembling the VW car model in which the Bosch starters will be used. In this way, the further reduction of man-hours is made possible for the vehicle manufacturer.

The prerequisite for adopting the modular production is to keep the production level high and steady. Furthermore, as modular production has to move finished component systems within the auto plant and build them together on production line as vehicles, its application on mini-car production is easier , better and more effective compared with on bigger vehicles.

Japan's Suzuki Motor, the world's biggest manufacturer of mini-vehicles, has been so successful in producing these vehicles at minimum cost that the more powerful manufacturers such as Toyota dared not to compete against it in mini-vehicle market.

A modular component system was originated by Osamu Suzuki, president of Suzuki Motor. For example, Suzuki has already had in-plant suppliers for a few years much earlier than when Lopez advocated such a system. President Suzuki commands rather than manages his company, somewhat as VW's

Piech does. But there is a difference between them. Although Piech recruited Lopez for the latter's capabilities, Suzuki himself is a "Lopez." He tells his men that the company must be different from Toyota or else it cannot survive, and that the company has to do things Toyota cannot afford to do because of its huge size or its pride.

Meanwhile, Daihatsu, one of Toyota's group companies, which had been a leading manufacturer of mini-vehicles from the 1960s through the 1970s, has had to settle for the number-two spot for more than 15 years. Toyota increased its holdings in Daihatsu to one third in September 1995. Daihatsu has been making every effort in their power to catch up and overtake Suzuki. The company has succeeded in increasing market share by introducing new products and in cutting cost with the resulting bigger profits.

In Europe, manufacturers of mini-cars such as Fiat, Renault, and local Big Three affiliates are reportedly trying to adopt modular production principles.

Until recently, most of the world's auto manufacturers have paid little attention to mini-vehicles because their added value is so limited that they are not profitable. But new concepts such as modular production might make greater profits possible, encouraging manufacturers to put more resources into mini-vehicle production. A mini-car has traditionally served as cheap, no-frills transportation. Manufacturers limit options and luxuries to keep mini-car prices low. But by succeeding in cost cutting, manufacturers might be able to add a bit more luxury to mini-cars in the future and the mini-car market may in fact grow. It is most probable that the renaissance of smaller- or mini-cars is coming. However, as manufacturers have to keep production levels high and steady for the cars to be economically feasible, fierce competition among them is unavoidable.

About the Author

Born in Mie Prefecture in 1924, Akira Kawahara graduated from Tokyo Imperial University in 1946 with a degree in mechanical engineering. After just one year in his first job with Mitsubishi Trading Company, he resigned to pursue his dream of working in the automobile industry. Because postwar Japan held few direct opportunities in automaking, Mr. Kawahara joined the Japanese Ministry of International Trade and Industry. He worked in the Ministry's Automobile Section for 10 years, helping formulate policies that fostered the regrowth of the Japanese auto industry. In 1957, Mr. Kawahara joined Toyota Motor Sales, Ltd. during its inaugural year of business in the United States, serving in a number of crucial positions, including Chief of Product Planning, member of the Board of Directors, and Managing Director. In 1981, he was appointed Senior Managing Director of Tokyo Toyopet Motor Sales Company. A year later, he was named Executive Vice President of Toyota Vista Tokyo Motor Sales Company, becoming president in 1983 and chairman in 1986. Mr. Kawahara concluded his distinguished career as an advisor to the Institute for International Economic Studies, and retired in June 1996.

Reference

Abernathy, William J. *Productivity Dilemma: Roadblock to Innovation in the Automobile Industry.* Johns Hopkins University Press, 1978.

Abernathy, William J., et al. *Productivity and Competitive Cost Advantage.* U.S. Department of Transportation, February 13, 1981.

Automobile Section, Ministry of Commerce and Industry. *Automobile Section News No. 11.* Automobile Section, Ministry of Commerce and Industry, July 1951.

Consumer Union. Annual Auto Issues of *Consumer Reports* 1970–1997.

Dertouzos, Michael L.; Lester, Richard K.; Solow, Robert. *Made in America—Regaining the Productive Edge.* The MIT Press, 1989.

Gabot, Andrea. *The Man Who Discovered Quality.* Times Books, 1990.

General Motors. General Motors' Annual Reports 1950–1996. General Motors, 1986.

Ingrassia, Paul; White, Joseph B. *Comeback—The Fall & Rise of the American Automobile Industry.* Simon & Schuster, 1994.

Inoue, Shoichi. *History of the American Automotive Industry (J).* Sanyo Tosho Shuppan, 1975.

J.D. Power and Associates. J.D. Power Surveys (see Index).

Japan Automobile Industry Federation. *History of the Japanese Automobile Industry (J).* Japan Automobile Industry Federation, 1979.

Japan Automobile Manufacturers Association. *The History of the Japanese Automobile Industry (J).* Japan Automobile Manufacturers' Association, 1988.

Jones, Daniel T.; Womack, James P.; Roos, Daniel. *The Machine That Changed the World.* Rawson Associates, 1990.

Kawahara, Akira; Nakajima, Atsuo; Sato, Ryuzo. *The Future of the Japanese Auto Industry (J).* Jidosha Shuuhou Sha, June 1957.

Kawashima, Takenobu. *Legal Consciousness of the Japanese (J).* Iwanami Book Co.

Keller, Maryann. *Collision—GM, Toyota, Volkswagon and the Race to Own the 21st Century.* Currency and Doubleday, 1993.

Keller, Maryann. *Rude Awakening—The Rise, Fall, and Struggle for Recovery of General Motors.* William Morrow and Company, Inc., 1989.

Krafcik, John F. *Triumph of the Lean Production System.* Sloan Management Review, Fall, 1989.

Lee, Albert. *Call Me Roger.* Contemporary Books, 1988.

McAlinden, Sean P., Andrea, David J.; Flynn, Michael S., Smith, Brett C. *U.S.-Japan Bilateral 1994 Automotive Trade Deficit.* Transportation Research Institute, University of Michigan.

Nakajima, Osamu. *The Road to the Lexus/Celsior (J).* Diamond Co., 1990.

Nader, Ralph. *Unsafe At Any Speed —The Designed-In Danger of the American Automobile.* Grossman Publishers, Inc., 1965.

Ohno, Taiichi. *Toyota Production System: Toward Effective Business Without Mass Production.* Productivity Press, 1988.

Petersen, Donald. *A Better Idea.* Houghton Mifflin, 1991.

Prestowitz, Clyde V., Jr.; Willen, Paul S. *Future of the Automobile Industry—It Can Compete, Can It Survive?* Economic Strategy Institute, 1992.

Sloan, Alfred P. *My Years With General Motors.* Doubleday, 1963.

Suzuki, Ichiro; Okamoto, Kazuo. *The Toyota Celsior (J).* Journal of SAE Japan No. 3, 1990.

Tobita, Hideaki; Arai, Masuji; Shibata, Youichi; Muda, Hirobumi. *Development of a Flexible Production System for Body Welding (J).* Journal of SAE Japan No. 1, 1991.

Toyoda, Eiji. *Fifty Years in Motion.* Kodansha International.

Toyota Motor Corporation. *The 50-Year History of Toyota Motor Corporation (J).* Toyota Motor Corporation, November, 1987.

Toyota Motor Sales Co. *The Thirty-Year History of Toyota Motor Sales Co. (J).* Toyota Motor Sales Co., December, 1980.

U.S. Trade Representative. *1994 Annual Report on Trade Barriers in Foreign Countries.* U.S. Trade Representative.

Yamazaki, Kiyoshi. *Automobile Power in Japan, U.S.A. and Europe (J).* Diamond Co., 1971.

Yates, Brock. *The Decline and Fall of the American Automobile Industry.* Empire Books.

Index

Index

MIX
Papier aus verantwortungsvollen Quellen
Paper from responsible sources
FSC® C105338

If you have any concerns about our products,
you can contact us on
ProductSafety@springernature.com

In case Publisher is established outside the EU,
the EU authorized representative is:
Springer Nature Customer Service Center GmbH
Europaplatz 3, 69115 Heidelberg, Germany

Printed by Libri Plureos GmbH
in Hamburg, Germany